Teaching and Learning Science

TEACHING AND LEARNING SCIENCE
A Handbook

VOLUME 1

Edited by Kenneth Tobin

PRAEGER

Westport, Connecticut
London

Library of Congress Cataloging-in-Publication Data

Teaching and learning science : a handbook / edited by Kenneth Tobin.
    p. cm.
    Includes bibliographical references and index.
    ISBN 0–313–33573–7 (set : alk. paper)—ISBN 0–313–33936–8 (v. 1 : alk. paper)—
ISBN 0–313–33937–6 (v. 2 : alk. paper)   1. Science—Study and teaching.
1. Tobin, Kenneth George, 1944–
    Q181.T3513   2006
    507.1'2—dc22   2006015088

British Library Cataloguing in Publication Data is available.

Library of Congress Catalog Card Number: 2006015088
ISBN: 0–313–33573–7 (set)
        0–313–33936–8 (vol. 1)
        0–313–33937–6 (vol. 2)

First published in 2006

Praeger Publishers, 88 Post Road West, Westport, CT 06881
An imprint of Greenwood Publishing Group, Inc.
www.praeger.com

Printed in the United States of America

The paper used in this book complies with the
Permanent Paper Standard issued by the National
Information Standards Organization (Z39.48–1984).

10 9 8 7 6 5 4 3 2 1

Contents

Preface

In many ways, life as a science teacher was simpler back in the 1960s, when I began to teach science in a small junior high in Western Australia. The space program was in full flight and man was yet to walk on the moon. Science offered so much, and as a physics teacher, my lab coat and slide rule were "badges of honor." Most teachers saw themselves as distant from science and people who knew about it and taught it. Charles Percy Snow had brought to our attention the two cultures, the sciences and the humanities. From the perspective of a science educator, this set and the associated debates were good reason to reach out to the masses who not only did not have a good understanding of science but also did not see themselves as being able or wanting to communicate with science types. Forty years later, the problems seem worse. The space program is on hold again because of safety problems with space shuttles, and yesterday a physicist testified before a federal judge in favor of science programs drawing attention to intelligent design as an alternative to evolutionary theory. The conflation of science and religion is not a new problem, but the ongoing debate and its fate being decided in the courts are troublesome. Recent polls suggest that more than 50 percent of Americans reject evolution as a viable theory, accepting some form of creationism and supporting the teaching of creation science in public schools. Added to this worrying trend is recent publicity about evangelical physicists arguing in favor of divine or intelligent falling as an alternative to gravity. I use these examples to support an assertion that the case for educating the public about science is as strong as it ever has been. How does a public as complex as ours make sense of the arguments of physicists arguing for intelligent design against evolution, offering alternative explanations of the same graphs used by scientists arguing in support of evolution? Needless to say, I was concerned when I heard a reporter on a public radio station conclude that she found some aspects of the argument for intelligent design compelling and a school board member from a town near Philadelphia claim that he was convinced by the testimony.

This handbook is written for teachers, parents, students, and members of the public. Its purpose is to bring core issues in science education to the attention of a broad audience so that science education curricula and public debate are informed by contemporary thought in science and science education. At a time when the advancement in science is rapid on many fronts, there are also policy initiatives such as the adoption of *National Science Education Standards* and the No Child Left Behind legislation that perturb the status quo regarding science education in K–12 education. Teachers face many dilemmas as they plan and enact science curricula, and so do parents and students who seek to educate themselves to be scientifically literate. The process of selecting authors for the chapters of this set was to identify a cadre of the world's leading science educators, from the United States and elsewhere in the world, inviting them to write for the target audience about the issues for which they are renowned. At the same time, we identified leading science educators involved in K–12 education but not affiliated directly with schools or universities, and science teachers who were known for their exemplary practices. Finally, in an effort to illustrate how the frontiers of science could be woven into the science curriculum, we were fortunate to gain the involvement of leading scientists who have also contributed to science education.

The sixty-six chapters of the handbook vary in length and focus. Some chapters address history of science and its place in the curriculum, others review research in science education and present the implications for consideration, and other chapters provide innovative suggestions for teaching and learning science. Each chapter underwent a process of rigorous peer review, and those accepted for publication were sometimes further revised to fit the goals of the set. The result is a set of chapters that covers what I regard as the central issues in science education in ways that will stimulate thought and catalyze changes in practice.

The set consists of eight parts, the first addressing pressing issues in science education and the last examining contemporary science in a context of its inclusion in the curriculum. In between are six clusters of chapters that investigate language and scientific literacy, home and school relationships, equity, new roles for teachers and students, making connections with science, and resources for teaching and learning science. The result is a unique collection of scholarly articles, written for the public, that has the potential to stimulate thinking, debate, and changed practices in K–12 science education, whether it is practiced at home, in informal settings, or in schools. The hope of the authors is to invite further thought and inquiry about the issues addressed through explorations of the myriad additional resources identified in the chapters and through ongoing investigations with the goal of bringing scientific literacy to the public at large so that the citizenry is not bewildered by contemporary science, technology, and their interfaces with society and political debate.

The final part, which addresses science in the news, consists of three sets of chapters. Chapters 56–58 are written by leading scientists who also have an active role in science education. The chapters address frontiers of science by leaders in their fields. A second set of chapters, Chapters 59–62, addresses science in the public interest, notably issues of health and the environment. Angela Calabrese Barton and a group of colleagues from Teachers College, Columbia University, developed the chapters. Similarly, Catherine Milne collaborated with a group of her graduate students to produce a set of chapters (Chapters 63–66) that addressed applications of chemistry in everyday life. Catherine describes the approach and purpose of these chapters in the next section.

EVERYDAY CHEMISTRY: AN INTRODUCTION

> At the end of the detailed chemistry core curriculum there is a content connections table that not only lists and describes all of the major understandings but also includes real world connections. However, of the 114 major understandings presented only 46 actually have any real world connections listed. These real world connections include spectral analysis of stars, metallurgy, electroplating, free radicals, degrees of saturation of solutions and the Haber process. Can students really make these connections to the real world? Wouldn't students have an easier time connecting to the real world if they looked at chemistry that was more relevant to their lives such as the chemistry of everyday things?
>
> —Kristin McNichol, 2005

Kristin McNichol, a former graduate student, asked these questions as she introduced her paper on whether the chemistry of everyday things had a place in chemistry curricula. It seems like a good place to begin my introduction of a series of papers written by students who wrote the first iteration of these papers as they were working toward their certification to teach science in middle and high schools.

Biology is often in the news. There are reports in the newspaper about the latest animal or plant genome to be unraveled, including that of the chimpanzee, one of our close cousins; the fruit fly, an important model organism for research; and the rat, a household pest. There are also reports about the cloning of dogs, frogs, and sheep; and debates about stem cell research. In physics there are reports in the newspaper of ongoing debates about the actual number of planets in the solar system, whether Pluto is actually a planet, and how we can understand the destructive power of hurricanes. Apart from the ongoing discussion of the dangers of trans fats, also called *partially hydrogenated oils*, chemistry seems to be pretty much ignored by comparison.

This was one of the reasons I encouraged students in a chemistry course for teachers to investigate cutting-edge developments or aspects of our everyday lives and to tell stories that revealed their chemistry. The authors tried to ensure that they structured their chapters to begin with phenomena that would be familiar to their reading audience of parents, teachers, and students. Beginning with phenomena is one way of beginning a process of understanding that connects our prior experiences with current understandings. Understanding also requires explanation, and, as you will note in the following chapters, explaining phenomena requires reference to objects such as atoms and molecules that make up all matter but cannot be observed directly, at least not by learners in school. The fact that these particles cannot be observed directly makes the learning of chemistry a challenge. For example, I can take two giant Wint-O-Green Lifesavers, go into a really dark room, wait for my eyes to adjust to the dark, and then bang the Lifesavers against each other. As I do this, if I am lucky and observant, I will see a small bright blue flash. This phenomenon is called *candy triboluminescence*. Try it for yourself! When I asked my chemistry class whether any of them had experimented with this phenomenon, one of the sixteen students indicated that she had experimented with Wint-O-Green Lifesavers when she was much younger. However, explaining that phenomenon is a bit more difficult. It requires an understanding of energy transfer, positive and negative charges, and excitation of electrons in nitrogen molecules that emit the blue light you observe as the electrons in the molecules release the energy they absorbed that began with the banging of the candies together.

This is the challenge facing the four authors in this part: how to present phenomena and then provide an explanation that is both scholarly and understandable. Danielle Dubno's examination of the nutrition "facts" that are required by law to be displayed on processed food items led her to an investigation of what the term *partially hydrogenated soybean oil* meant and resulted in a more detailed examination of the chemistry of the synthesis of trans fatty acids and the implications of their use for human health (Chapter 63). Danielle's study of trans fats emphasizes the important relationship between the structure of molecules and their function, and raises the questions of how our bodies respond to synthetic molecules and what the implications of these responses are to human health. Marsha Smith was inspired by the students with whom she worked and by her personal experience to write her chapter (Chapter 64) on the chemistry of hair relaxers. Almost everyone has hair, and this provided a starting point for Marsha's examination of the chemical processes involved in manipulating the alkalinity and acidity of hair to control the breaking and formation of strong covalent bonds that determine the curliness or straightness of hair. Ji-Myung Nam describes an investigation using sunprint paper to examine the relationship between sunblocking potential of sunburn lotions and their skin protection factor (SPF) and the thickness of the applied sunscreen (Chapter 65). We highlight the use of questions for inquiry and how you might develop an investigation to explore the questions using science processes. Although not reported in her chapter, Ji-Myung confirmed through studies of the effect of sunblocks and sunscreens and her own skin that sunprint paper works well as a model for human skin. In Chapter 66, Sophie Homan brings her background in physics and her concern for the environment to bear on her examination of the chemistry of hydrogen fuel cells and the technical, environmental, and economic challenges facing the developers of usable hydrogen fuel technology. Her examination of hydrogen fuel cells showed how the fuel cell structure separated the oxidation and reduction reactions of combustion so that the energy produced could be made available to do work such as powering an automobile. Sophie's study seems particularly prescient as the price of gasoline continues to rise and alternative energy sources become more attractive.

These chapters challenge science educators to reflect on the relationship between phenomena that are experienced and explanations that require the use of molecules and atoms, and to ask themselves how they make those connections when they work with learners trying to make sense of science. These chapters also highlight the chemistry in everyday things and ask why everyday items do not make a more significant contribution to current chemistry curricula. As you read these chapters, I invite you to consider whether learners might enjoy chemistry more if they had the chance to in-vestigate and learn the chemistry of new resources or everyday items similar to those that form the focus of these chapters.

Kenneth Tobin

Acknowledgments

The research in this set is supported in part by the National Science Foundation under Grant No. DUE-0427570. As part of a program of research supported by a Distinguished Teaching Scholar award, I undertook dissemination of research, scholarly thought, and exemplary practice in science education. This publication is part of this dissemination effort to a public that extends beyond academics in universities. Any opinions, findings, and conclusions or recommendations expressed in the book are those of the authors and do not necessarily reflect the views of the National Science Foundation.

Catherine Milne and Kathryn Scantlebury were of considerable help in identifying suitable authors for this handbook and, in Catherine's case, in editing and coauthoring a number of the chapters. Their professional competence, enthusiasm, and optimism greatly contributed to the success of this venture and yielded a first-class set of cutting-edge scientists and science educators.

I greatly appreciate the ongoing support of Barbara Tobin, who patiently watched the various deadlines whizz by with understanding and good humor, having produced her own chapter well ahead of the due date.

Finally, and surely not least, I thank Wolff-Michael Roth for his friendship, support, and suggestions throughout this entire project.

PART 1

KEY ISSUES IN
SCIENCE EDUCATION

1

Analyses of Current Trends and Practices in Science Education

Kenneth Tobin

Our Nation is at risk. Our once unchallenged preeminence in commerce, industry, science, and technological innovation is being overtaken by competitors throughout the world. . . . What was unimaginable a generation ago has begun to occur–others are matching and surpassing our educational attainments.
—National Commission on Excellence in Education
(1983, opening paragraph)

One way or another, international comparisons have driven policy in regard to science education, at least since the launch of the Sputnik satellites in the 1950s and probably long before that. Possibly it is the links between science, technology, and military prowess that have fueled such comparisons, and perhaps underlying all concerns is a goal to retain economic competitiveness with all nations of the globe. Watershed events in science education have been associated with such phenomena as Sputnik, *A Nation at Risk* (National Commission on Excellence in Education 1983), and the No Child Left Behind Act (U.S. Department of Education 2001). As George DeBoer (1991) points out in his analysis of the history of science education, events such as these are catalysts for trends that may already have been in the offing. For example, a focus on the structure of the discipline, following Sputnik, was associated with a rising dissatisfaction with progressive forms of education that had characterized the first half of the twentieth century. The launch of Sputnik raised fears about the loss of military supremacy and, in a context of the Cold War that followed the Second World War, education concerns intersected with worries about the adequacy of the pipeline for producing future generations of scientists and science, which would presumably accelerate technology for economic and military prowess. DeBoer notes somewhat ironically that the concerns of progressivists, such as John Dewey, for child-centered pedagogy and curricular relevance were swept aside with science curricula that focused on understanding science and stripped away applications and efforts to build curricula around students' interests. The irony in this observation is that nearly a half-century after Sputnik, the failure of enacted science curricula to align with such criteria as relevance and interest may underlie many of the pervasive and continuing equity issues.

SCIENCE EDUCATION AND THE NATIONAL INTEREST

In the decades that followed Sputnik, the focus in science education was on curriculum development, with a plethora of curriculum materials being developed with the assistance of federal and private resources. A fairly pervasive shared goal was that students would learn science through inquiry, by doing science, and through the use of process skills based on analyses of what scientists did while doing science. Collaboration of scientists with educators and psychologists led to a gradual breakaway from behaviorism as the psychologies of Jean Piaget, Jerome Bruner, and David Ausubel were incorporated into curriculum resources developed with input from practicing scientists, such as Robert Karplus, in collaboration with educators. Although many have claimed that the curriculum development efforts of the 1960s aimed to teacher-proof science education, many of the efforts, such as the Elementary Science Study, were open-ended and reliant on excellent teaching. In fact, at each level of schooling, the available curriculum materials ranged in their pedagogical approach from highly structured approaches, such as in *Science— A Process Approach* (American Association for the Advancement of Science 1967), to very open-ended approaches, anticipating that students would create structures from their experiences, interests, and enjoyment. The open-ended approach is described by David Hawkins (1965) in his classic article "Messing about in Science," and is incorporated into the Elementary Science Study. However, no matter what approach curriculum developers suggested in their materials and which project school developers adopted, the ways in which science education were enacted were greatly dependent on the teachers and students. The interactions between teachers and students were as central to the creation and maintenance of productive learning environments as curriculum resources and materials. In Chapter 7, Barry Fraser explores a program of research on learning environments in science education that has spanned several decades.

In science education, there were numerous focused programs of research on learning to teach science, approaches that paralleled the development of new curricula. For example, Mary Budd Rowe's (1974) famous studies on wait time emanated from a concern that, even though large sums of money were being directed toward the development of inquiry-oriented science curricula, such as the Science Curriculum Improvement Study, indicators of inquiry were infrequently observed and did not reflect the curriculum materials used by the teacher. In particular, Rowe was puzzled by a lack of inquiry in the verbal interactions between teachers and students, and studied a number of tape recordings of science lessons. From about 200 tapes, she identified less than a handful that showed evidence of verbal inquiry. These, she realized, incorporated relatively lengthy pauses (i.e., wait time) between successive speakers. From these analyses, Rowe initiated a program of research in which teachers and students manipulated the average wait time, finding that when the average wait time exceeded three seconds, there was widespread evidence of verbal inquiry and science achievement increased.[1] For example, when teachers paused for more than three seconds after asking a question, students responded more often, more appropriately, and at greater length. Also, instead of responding, students sometimes asked questions of their own and showed other evidence of thinking about science. Not surprisingly, teachers changed their expectations of what students could do, especially those they had previously regarded as low achieving.

The energy crisis, inflation, and concerns about the quality of the U.S. military were factors often associated with the science education budget at the National Science Foundation (NSF) being reduced almost to zero in Ronald Reagan's initial years as president. With a severe reduction in federal resources, the science education community seemed destined for lean years. However, a catalyst for renewed funding was the *A Nation at Risk* report quoted at the beginning of this chapter (National Commission on Excellence in Education 1983). Publication of the report drew the attention of the nation and the world to an asserted connection between the quality of a nation's education and its commercial, industrial, scientific, and technological health. From the date of publication of the report and throughout the 1980s, the National Science Foundation budgets for science education gradually rose and once again supported the activities of science educators—such as curriculum development, uses of technology, teacher preparation, professional development, and, eventually, research. As the decade of the 1980s unfolded, the growth of Japan's economy and the demise of the Soviet Union shifted the focus of science education from producing scientists and engineers for military purposes to maintaining an economic edge and international competitiveness. The Soviet Union declined as an economic and political power, the Cold War ended, and economic rationale bolstered support for enhancing science education in the United States.

Concerns about international competitiveness were fueled initially by Sputnik, leading to the first international study of science education in the mid-1960s (National Center for Education Statistics 1992) and subsequent waves of international testing culminating in 1995 with the Third International Mathematics and Science Study (TIMSS; n.d.). The results showed that U.S. students lagged far behind their Asian counterparts, creating an argument for focused attention on improving the quality of science education, with the goal of moving up the rankings to be number one. Within the United States, the goal was supported by a plethora of state- and national-level efforts to improve quality through testing and accountability measures. For example, when I came to Florida in 1987, I was appointed to a Commissioner of Education's Task Force to create a comprehensive plan for science and mathematics education. The State Chamber of Commerce and Florida's Department of Education jointly sponsored the task force. From the outset, the intention of the task force was to create a plan that ensured excellent science achievement for all Floridians and high levels of accountability through statewide testing. Similar reform efforts occurred in most other states, and the NSF and U.S. Department of Education also provided resources to support the development of K–12 standards that were systemic across a state.

The 1990 publication of *Science for All Americans* (Rutherford and Ahlgren 1990; see also American Association for the Advancement of Science 2006) became a roadmap for many states and subsequent projects seeking to reform science education systemically. James Rutherford, who had previously written about the goals for science education in terms of conceptual schemes, took a lead role in developing a book that had a revolutionary impact on science curriculum reform since it was incorporated almost entirely into the *National Science Education Standards* (National Research Council 1995, 1996). One of the central aspects of *Science for All Americans* is often referred to as "Less is more." The rapidly expanding knowledge base of science could be represented in terms of a set of central themes, such as conservation of energy. Once identified, these themes could be learned through investigations of many topics. Rather than attempt to cover all, or even most, topics that have traditionally been covered in a subject such as chemistry, the premise is that for any particular theme (or standard from the *National Science Education Standards*), it was beneficial to study a small, representative set of topics in detail, rather than to superficially cover many topics. In Chapter 4 of the handbook, Richard Kozoll and Margery Osborne explore meaning in science, examining issues of relevance and the lifeworlds of learners.

WHAT COUNTS AS RESEARCH?

Policy makers expected an alignment in national standards, curricula, assessment, and science teacher education. The No Child Left Behind Act of 2001 (NCLB; U.S. Department of Education 2001) is built on four principles: accountability for results, more choices for parents, greater local control and flexibility, and an emphasis on doing what works, based on scientific research. Since science was not one of the subject areas to be assessed when the act was first enacted, the immediate impact on elementary science was negative—the focus on reading and mathematics was so great that science frequently was relegated to being taught late in the day or scarcely. Now with science to be among the subjects assessed every year, beginning in 2007, it is anticipated that the neglect it experienced at the elementary levels will vanish. However, other problems are likely to emerge. For example, the curriculum across the grade K–12 spectrum may focus on what is tested, with consequences for what is learned and how it is learned. Furthermore, the ways in which science is to be taught—that is, what works—are to be guided by scientific research, as defined by policy makers. Is this yet another way to teacher-proof the curriculum? Or, worse, is it a way to standardize what is taught in teacher preparation and professional development programs? Already efforts have been made to filter the massive research in science education to retain those results that are based on some forms of inquiry and to erase others that were not produced "scientifically" (Shavelson and Towne 2002). Just how is scientific research in the social sciences defined?

Richard Shavelson and Lisa Towne produced a report of the Committee on Scientific Principles for Education Research (2002) in which different forms of inquiry were identified. The report is a well-balanced analysis of forms of inquiry and associated quality criteria. However, the No Child Left Behind Act defines the term *scientifically based research* as "research that involves the application of rigorous, systematic, and objective procedures to obtain reliable and valid knowledge relevant to education activities and programs." For research to be regarded as scientific, it must employ systematic, empirical methods; use measurements or observational methods; undertake rigorous data analyses; and evaluate outcomes using experimental or quasi-experimental designs. Replication and peer review are also regarded as central features of scientific inquiry.

Policy initiatives, such as No Child Left Behind, have enormous ramifications for science education because of their impact on funding of research and development. Specification of the criteria for scientific inquiry is welcome in the sense that scholars should continuously question the viability of their methodologies. What is troublesome in the instance of the ripple effects of No Child Left Behind is that one perspective seems to hold sway, one that I regard as regressive. I embrace what Joe Kincheloe (2005) has so often written about as a bricolage model, whereby multiple framings of social phenomena inform scholarly inquiry. Hence, I can see the relevance, in some circumstances, of using quasi-experiments to explore different approaches built on programs of research. For example, emerging from our research on coteaching is an approach to improving science education that we call *cogenerative dialogue.* Sarah-Kate LaVan discusses cogenerative dialogue in Chapter 31, so here I simply use it as an example. Cogenerative dialogues are discussions between teachers and (usually) students about shared experiences in a classroom. Their purpose is to identify and resolve contradictions in enacted curricula and in so doing develop a collective responsibility for agreed-to changes in roles, rule structures, and resource use in a science classroom. It turns out that cogenerative dialogues are also very useful as seedbeds for teachers and students to learn about one another and especially how to interact successfully across cultural boundaries that are created in part by differences between the teacher and students by factors such as age, social class, race, and gender. Our knowledge of the potential of cogenerative dialogues and how they might be applied in science classrooms emerged from years of careful interpretive research focusing on teaching and learning and on how to teach in urban schools. At the beginning, we did not expect that cogenerative dialogues would be a potentially powerful intervention in subjects other than science or beyond urban high schools in the United States. Having done the research and development, it now makes sense to me for a quasi-experiment to be planned to ascertain if, on a scale of many classes in many schools, the uses of cogenerative dialogues lead to valued changes in teaching and learning, including the production of higher achievement.

Having fought so many years to gain acceptance for interpretive forms of inquiry that employ qualitative data and participatory methods to explore patterns and associated contradictions, it is a body blow to see legislation that apparently ignores the importance of these forms of inquiry and assigns greater weight to forms of research that align with the stereotypes of research in the hard sciences. Although I acknowledge that the wording of the act does not preclude interpretive research, funding agencies and program officers may take a perspective that there is greater payoff in supporting experiments and quasi-experiments in science education. My fear is that such research will not only support the status quo but also lead to forms of prescription that will not work across social and cultural borders. It would be regressive if forms of inquiry that produce exciting new directions in science education were stifled, such as Rowe's research on wait time and our interpretive studies on cogenerative dialogues.

EXPANDING THE PURPOSES OF SCIENCE EDUCATION

Just what does it mean to do and learn science? Presently we have too narrow a focus. For example, in New York City, the gold standard is attaining a good score on the Regents' Examinations in the different sciences. I do not have a problem with students needing to show that they can pass a Regents' exam at an appropriately high level, but I do have big problems with curricula focusing on this as the raison d'être for science. Enacted curricula can be impoverished indeed when teachers do a task analysis of past examinations, teach to prepare students to answer questions like those asked in the past, and take the time each week to teach students how to take tests. I do not dispute that doing well on the Regents' Examinations counts a great deal in terms of opening doors for some students, but here I raise the possibility that the consequences of focusing solely on passing examinations might have dire consequences for students who may have very little interest in and use for science, and may also have such consequences for the nation, which is already struggling to find candidates for science and engineering studies and employment. In Chapter 2, William Boone and Lisa Donnelly explore some aspects of high-stakes testing.

Science, and science-related domains such as engineering and technology, can be thought of as being enacted in numerous fields in society. For example, science is in the newspaper, on television, in the home, in leisure activities, at the shopping center, in workplace sites, and in institutions like hospitals, parks, museums, fire departments, and zoos. Students might identify some of these fields to reflect their interests and priorities on what is important to know. Personally, I access science daily on the Internet, often going to popular news and professional sites like CNN, BBC, AAAS, NSTA, and NASA. Also, Wikipedia (n.d.), an open-access encyclopedia, is a burgeoning resource from which students can learn and also contribute. At sites such as these, I read stories for the public and become informed about science and its connections with so many aspects of humanity—including weapons, health, origins of man, sports, and frontiers such as nanotechnology. I cannot imagine any science education program that does not contain a component that explores relevant science on the Internet as it unfolds from day to day. If the course were chemistry, then students would search favorite sites for contemporary chemistry. Not only is the Internet a wonderful resource, but it is also a source of misinformation. Hence, an important skill is to test for false claims by cross-checking accuracy using multiple sources. This is an important skill to have not just in high-tech fields like the Internet, but also in social life generally.

I regard science as a form of culture, and as such, it is important that students have opportunities to do science—not only to see its relevance, but also to learn science in ways that are potentially transferable to life outside of school. Much of what is learned through curricula enactment in many fields will reproduce canonical science, as it is represented in the *National Science Education Standards*, and in other cases students will transform what they learn to meet their own goals. For example, a student tuning into CNN this morning would have experienced a compelling program on food allergies. The program featured CNN's senior medical correspondent, Dr. Sanjay Gupta, and his guest, a medical researcher with a huge federal grant to support the development of a vaccine for food allergies. The program had immediate import for the everyday practices of students and was informative in terms of the science of allergies and how to change lifestyles to avoid health risks. Students who watched the tape and subsequently explored other re-sources, such as Internet sites, might represent what they have learned as changed diets and improved health. If, after watching the program, students visited the CNN health website (CNN.com; n.d.), they would have the option of reading the latest news on the approval of brain stem cell transplants, new breast cancer drugs, policy moves in Europe in response to the appearance of bird flu virus, and an investigation into whether or not antibacterial soaps work. Reading current reports such as these can contribute to students building interests in science, changing their lifestyles, and building a science-related vocabulary they can use if and when the chance arises in their interactions with others and resources such as the media.

Just as it always has been, a central part of an effective science education will involve students learning facts and procedures for solving the kinds of problems scientists do. A key is that students learn a basic core of facts so that they can be fluent in their use of science language when they interact with others. In chemistry, for example, it seems essential that students know basic facts such as the symbols for the common elements and their place on the periodic table, valence, and atomic mass. I do not mean that the students should only know how to look up facts or retrieve them from a periodic table or textbook (though they should know how to do these things); I expect them to be able to reproduce facts rapidly, on demand, and without having to go through a conscious process. That is, if students were to hear the word *oxygen*, many of the properties of oxygen would be available as a conversation unfolded, allowing them to make sense of the conversation and, as the opportunities arose, to contribute to it. Knowing the approximate atomic weights of such elements as carbon, oxygen, hydrogen, sulfur, nitrogen, calcium, and magnesium would be part of a basic science toolkit, factual knowledge that could be recalled quickly and used in interactions in and out of science class. Similarly, I expect students to know formulae for such compounds as calcium carbonate, sulfuric acid, sodium chloride, and sodium carbonate and to specify their approximate formula weights within a few seconds. Of course, being able to look up such facts also is part of an essential science toolkit, but having to stop and look them up disrupts fluency and the likelihood that science can be used outside of the classroom when lookup resources are not conveniently available.

A test for science education is whether or not a person is able to enact science to appropriate the resources of a field to meet her goals. Often this means using science appropriately in a just-in-time way, usually in an anticipatory manner—without having to stop and think. Situations in which science might be used fluently include being a literate citizen, making sense of hot political debates such as the relationships between hurricane activity and the incidence of greenhouse gases, federal funding for stem cell research, and efforts to teach intelligent design in public schools. Similarly, science can play a role in using the technology of the home without being intimidated by it—from the mundane, such as a refrigerator and thermostat, to the more exotic such as the large-screen TV and wireless Internet.

I am not opposed to students becoming savvy about test taking, just on schools focusing a curriculum too narrowly on what is usually tested. Among the ways in which students can show what they know is to do test items on a regular basis. In a given week at school, I'd very much like to see students participate in a wide variety of fields in which science is enacted. Of course, the curriculum should be planned and enacted purposefully. However, perhaps students can be involved in the planning of what is done, how it is done, and when it is done. Rather than putting the focus on test items only, the solving of test items might be done for just a short period of time several days a week. Similarly, within a context of less is more, science in newspapers, magazines, and books could be explored for a short periods several times a week, and important facts can be memorized as part of a science toolkit. A well-balanced science program would use different ways of organizing students for learning, including whole-class interactions, small-group collaboration, and individualized forms of participation. The outcomes of science education would extend beyond students developing interests in science and getting high test scores, to include students being able to talk, write, and read fluently about what they have learned. I see it as a high priority to set up classes so that all students get to tell others what they learned and what they expect to pursue next. Written and iconic forms of representing what has been learned and what is known also should complement oral uses of language.

NEW WAYS TO CONSIDER LEARNING

In the wake of the Sputnik curriculum explosion, the theoretical underpinnings gradually evolved from behaviorism to conceptual change models of learning grounded in social constructivism. Joseph Novak was one of the leading thinkers that moved the field in this way with his 1977 book and numerous publications. In Chapter 3, David Treagust describes science education within the conceptual change framework. As is the case with all theoretical frameworks, they illuminate some key aspects of learning and knowing and obscure others. Hence, after almost three decades in which conceptual change models have dominated research and curriculum development, sociocultural models can provide insights into persistent problems, identify new contradictions, and create fresh implications for science education, including research, teacher education, and curriculum design.

When it is viewed from the perspective of science as culture, learning science is regarded as cultural production. The emphasis is on the fields in which science is enacted and their structures, that is, within a given field, the resources available to support the production and enactment of science culture. Within a field, cultural production and enactment are purposeful, though not always conscious, to participate in ways that contribute to a community reaching its collective goals and to afford the participation of others. From this perspective, there is more than knowing what, how, and when; actions must happen just in time, be appropriate, and be anticipated by others—thereby expanding others' agencies, or power to act appropriately in the field. Individuals pursue goals, but always in relation to collective goals—each person having a part to contribute to the success of the community. As participants enact roles, they do so in ways that contribute to a flow of cultural enactment, their practices being enacted without having to take time out to deliberate on what is appropriate and how to do what must be done. Although the time metrics vary from field to field, it seems important to be able to act when necessary without have to take a time-out and thereby disrupt the flow. This implies that students have to learn science by doing science in a number of fields, initially participating in a peripheral sense and over time building expertise and becoming more central.

The relationship between the agency of participants and the structure of a field is dialectical, that is, agency and structure are interdependent and cannot be considered separately from one another. The practices of individuals within a field are part of the structure and therefore become resources to be appropriated (i.e., used by participants to meet their goals) by all participants. When any person in a field acts, the action is a resource that can be used by the actor and all others within a community. Obviously, then, a person's interactions with the resources of a field are an important dimension of a learning environment, since they are part of structure—dialectically inter-connected with the agencies of all participants. It is never enough to consider a person's interactions only in relation to her own learning since what is done, when it is done, and whether or not what is done can be taken up by others as part of the structure of a field—thereby expanding or diminishing opportu-nities of all participants to learn.

Agency also concerns the capital that learners have. Cultural capital in-cludes culture that learners have produced in their school studies and in society, in fields such as leisure, the home, and the neighborhood. Other forms of capital are symbolic and social. Symbolic capital concerns status and membership symbols or acknowledgments—such as being smart, interested, and talented. Social capital relates to networks that allow for access to human and material resources, such as who a person knows and can successfully interact with. During interactions in a field, there is an exchange of capital that can allow for the production of culture—that is, as capital is exchanged, new capital is produced in what is not a zero-sum game. Hence it seems critical to set up a classroom field in ways such that all participants can get involved in ways that maximize their uses of capital and allow them to pro-duce new culture that can potentially expand the agencies of the collective. In our recent research, we have learned that participating in successful interac-tions produces positive emotional energy and allows for chains of successful interactions to spread throughout a field (Tobin, Elmesky and Seiler 2005). Hence, a worthwhile focus for teachers and students can be on arranging material and human resources in ways that maximize success throughout a community and minimize the number and duration of unsuccessful interac-tions. Accordingly, the sociocultural focus I have brought to this section of the chapter emphasizes the distribution of resources and their appropriation in interactions. A sociocultural perspective is concerned with interactions and the necessity to maximize success and the production of positive emotional energy. Positive emotions, to be maximized in a science class, include inter-est, enjoyment, happiness, and feelings of accomplishment and being rec-ognized. Conversely, negative emotions, to be avoided or minimized, include anxiety, frustration, annoyance, anger, resentment, sorrow, and feelings of being disrespected. The use of emotions and emotional energy as a barometer for the quality of classroom environments, in conjunction with the quality of interactions, suggests a pathway for improving science education.

EQUITY IN SCIENCE EDUCATION

Equity issues have long been a concern of scholars like Jane Butler Kahle (2004), who examined the participation and achievement of females in science in and out of school. Due in large part to the scholarly activities of science educators like Kahle, the failure of females to perform at the same level as males and to participate in science, especially the physical sciences, has been ameliorated. However, even today, gender-related issues remain an active concern of many science educators, especially in relation to the participation of females. Indeed, the provocative title of Kahle's 2004 article asks the question about whether girls will be left behind in science education. Kathryn Scantlebury addresses gender equity in Chapter 23.

From the sociocultural perspective I laid out in the previous section, equity issues can be couched in terms of whether or not students from particular groups are able to exercise their agency in a given field. Can students appropriate the structures to meet their own goals and contribute to the accomplishment of collective goals? With the focus on interactions with others and material structures, equity can be focused on whether or not they participate in successful interactions and generate positive emotional energy. In our ongoing research, we have found that teachers and students in urban schools have significant difficulties in successfully interacting across boundaries defined by gender, social class, race, and age. We have found it fruitful to explore these difficulties in the extent to which students and teachers can recognize one another's capital and appropriate it fluently. In essence, a lack of understanding of different forms of culture make it difficult to experience successful interactions and thereby to achieve individual and collective goals. Randy Yerrick has observed similar patterns and describes some of them in Chapter 20, in a study that is set in a low-track science class at a rural high school.

In dealing with equity, it is well to keep in mind the relationships between agency and structure. What forms of structure will expand the agencies of all participants and afford their participation and success in science classes? What appeals to me most about this approach is the promise of building enacted curricula around the interactions that unfold in a classroom, on connecting teaching practices to the capital of the students in ways that yield active participation and hence achievement. The focus on collective agency does not negate seeking to obtain additional resources to support appropriate forms of practice that would otherwise be difficult (i.e., increasing the number of high-speed Internet connections). Paying attention to collective agency has the promise of expanding agency and thereby affording practices that are science related, are successful, and have the potential to alter the identities of the youth participants. Also, such an emphasis breaks from the determinism inherent in many policy exhortations that assume that because class and race play a part in achievement gaps in subject areas like science and mathematics, the best (and perhaps only) solutions are economic. Our research suggests that the best solutions will be cultural—teachers and students must learn to interact successfully across boundaries of race, class, gender, and age, and, until they do, resources can be available and accessible, but fail to be appropriated in ways that lead to higher science achievement. A breakthrough in our research suggests that cogenerative dialogues may be seedbeds for the growth of new culture across boundaries that appear to define inequities.

Our initial foray into cogenerative dialogues was in conjunction with coteaching. The purpose was to get input from students in the class on aspects of the quality of teaching, and especially on "how to better teach students like me." We set up the cogenerative dialogues to include two coteachers and two to three students from a class. Often the groups also included university researchers and/or teacher educators. The students were selected from those who were most challenging and to have as much diversity as possible between them. We established a rule structure that as long as all speakers maintained respect for one another and did not take one another's turns, this was a place where it was fine to speak openly without fear of reprisal. The goal was to improve the quality of teaching and learning, and it was acknowledged that it would therefore be necessary to identify contradictions that arose and discuss how roles, rules, and resource availability and distribution might be changed. The responsibility for what happened in the class was not assumed to be the teacher's only, and all participants in the cogenerative dialogue had a shared responsibility for enacting agreed-to changes in the next lesson. Finally, participants in cogenerative dialogues had to be active listeners, share turns at talk, and ensure that previously discussed issues have been resolved satisfactorily before a new topic is introduced.

Cogenerative dialogues made a difference. First we noticed that changes agreed to in cogenerative dialogues were enacted in the classroom, with positive changes discernible over a period of time. Second, we noticed that within the cogenerative dialogues, an esprit de corps gradually emerged and the interactions between the participants became fluent and focused on collective goals. Since the students and teachers usually differed in age, race, and class, the successful interactions we observed in cogenerative dialogues were precisely what we wanted to see occurring in science classes. Having made this observation, we have paid attention to other settings in which teachers and students interact and now see that one-on-one tutoring can be regarded as a form of cogenerative dialogue in which culture can be produced by both participants— potentially available to be enacted in science classrooms. Although more research is needed before vouching for cogenerative dialogues as essential components of professional development programs, teacher preparation, and ongoing formative evaluation, we have seen enough already to encourage teachers to set them up and study how they benefit enacted curricula.

LOOKING AHEAD

The sixty-six chapters that comprise this handbook of science education are arranged into eight clusters, or themes. This first cluster addresses key issues in science education. Already I have alluded to many of these issues in this introductory chapter. Among the issues addressed are high-stakes testing, conceptual change theory, meaning in science, controversy over evolution, learning from laboratory activities, and learning environments in science classrooms. The subsequent clusters of chapters address language and scientific literacy, home and school relationships, equity, new roles for teachers and students, making connections with science, resources for teaching and learning science, and science in the news. The chapters vary in length and approach. Some are written by leading researchers in science education, others are written by eminent scientists, and exemplary practitioners from classrooms and informal science centers such as museums wrote many of the chapters. As a resource, this handbook represents a powerful collection of chapters that are at the leading edge of science education.

NOTE

1. See my 1987 review of research on wait time (Tobin 1987).

REFERENCES

American Association for the Advancement of Science (AAAS). 1967. *Science—A Process Approach*. Lexington, MA: Ginn.
———. 2006. About Project 2061. www.project2061.org/about/default.htm.
CNN.com. N.d. Health. www.cnn.com/HEALTH/.
Committee on Scientific Principles for Education Research, Richard J. Shavelson and Lisa Towne, eds. 2002. *Scientific Research in Education*. Washington, DC: National Research Council.
DeBoer, George. 1991. *A History of Ideas in Science Education*. New York: Teachers College Press.
Hawkins, David. 1965. Messing about in Science. *Science and Children* 2 (5): 5–9.
Kahle, Jane B. 2004. Will Girls Be Left Behind? Gender Differences and Accountability. *Journal of Research in Science Teaching* 41:961–69.
Kincheloe, Joe L. 2005. On to the Next Level: Continuing the Conceptualization of the Bricolage. *Qualitative Inquiry* 11:323–50.
National Center for Education Statistics. 1992. International Mathematics and Science Assessments: What Have We Learned? January. http://nces.ed.gov/pubs92/web/92011.asp.
National Commission on Excellence in Education. 1983. *A Nation At Risk: The Imperative For Educational Reform*. A Report to the Nation and the Secretary of Education. Washington, DC: U.S. Department of Education.
National Research Council. 1995. National Science Education Standards: Contents. www.nap.edu/readingroom/books/nses/html/.

————. 1996. *National Science Education Standards.* Washington, DC: National Academy Press.

Novak, Joseph D. 1977. *A Theory of Education.* Ithaca, NY: Cornell University Press.

Rowe, Mary B. 1974. Wait-Time and Rewards as Instructional Variables, their Influence on Language, Logic and Fate Control: Part One—Wait-Time. *Journal of Research in Science Teaching* 11:263–79.

Rutherford, F. James, and Andrew Ahlgren. 1990. *Science for all Americans.* New York: Oxford University Press.

Third International Mathematics and Science Study. N.d. Third International Mathematics and Science Study (TIMSS): 1995. http://isc.bc.edu/timss1995.html.

Tobin, Kenneth G. 1987. The Role of Wait Time in Higher Cognitive Level Learning. *Review of Educational Research* 57:69–95.

Tobin, Kenneth, Rowhea Elmesky, and Gale Seiler, eds. 2005. *Improving Urban Science Education: New Roles for Teachers, Students and Researchers.* New York: Rowman & Littlefield.

U.S. Department of Education. 2004. Public Law print of PL 107–110, the No Child Left Behind Act of 2001. www.ed.gov/policy/elsec/leg/esea02/index.html.

Wikipedia. N.d. Wikipedia. www.wikipedia.org.

ADDITIONAL RESOURCES

Suggested Readings

Dewey, John. 1902. *The Child and the Curriculum.* Chicago: University of Chicago Press.

Roth, Wolff-Michael, and Kenneth Tobin, eds. 2005. *Teaching Together, Learning Together.* New York: Peter Lang.

Websites

AAAS: www.aaas.org
BBC: http://news.bbc.co.uk
CNN: www.cnn.com
NASA: www.nasa.gov
NSTA: www.nsta.org

2

High-Stakes Science Testing

William J. Boone and Lisa A. Donnelly

A science test is administered to elementary students throughout a state. Often weeks are spent preparing for the test. Teachers bring food and drinks so that all students have a full stomach before the testing begins. When the test results are distributed, principals quickly look at how their school performed. The students of which teachers earned high science scores? Which classrooms did not earn high scores? Did the school improve in comparison to the last test? Many people review the test results published in the local newspaper. Those preparing to sell a home look at the test results for schools serving their neighborhood. Parents review the graphs and numbers summarizing the test results of their child's school. The events described above take place every year when a high-stakes science test is administered and publicized. But how do we best understand such tests? In this chapter, we offer a primer on high-stakes science testing by considering some hows, whats, and whys of such tests. We provide a user-friendly overview of the issue of high-stakes science testing, so that whatever your needs and interests are, you are provided with guidance. The shorthand terms for specific tests and initiatives may change with each passing year, but the topic of high-stakes science testing seems surely to be a topic of interest for many years to come. With the changing world economy, it seems likely that high-stakes science tests will continue as technology and science jobs are increasingly coveted and policy makers attempt to optimize the science learning of students.

WHAT ARE HIGH-STAKES TESTS?
WHY ARE SUCH TESTS IMPORTANT?

High-stakes tests are those used to make critical decisions and assessments about individuals and groups of individuals. Many different kinds of high-stakes decisions are made from such tests. Individual schools and teachers may use high-stakes science testing information, in part, to determine student placement in honors, special education, or remediation science classes. Schools may use student science scores on these tests to determine student admission, promotion, or graduation eligibility. Schools may also use high-stakes science tests to determine science teacher and science program effectiveness. Teacher hiring, professional development, and retention often depend upon students' test scores. Furthermore, individual teachers may use testing information to revise and adapt science instruction for students. Families may use student test scores to guide summer school planning and enrichment activities. At the state level, student science test scores can impact school accreditation, ranking, and funding. At the national level, high-stakes tests are used to compare each state's performance and to assess systemic reform efforts. Federal funding levels of school districts may be impacted by high-stakes science scores. High-stakes tests, whatever their forms, impact the life of someone or something.

TESTING FORMAT

No high-stakes science test is identical to any other high-stakes test. High-stakes science tests may take the form of a traditional multiple-choice exam that requires students to bubble in correct answers on an answer sheet. Many high-stakes science exams also employ problem and essay questions that require students to write out their answers. Some high-stakes tests like the Third International Mathematics and Science Study (TIMSS) include a performance component during which individual students attempt to solve a problem with simple science equipment such as a beaker, a balance, or a timer. In these cases, students may complete an answer sheet and/or be observed by trained "judges" who evaluate the student.

Students complete high-stakes tests in many different testing environments such as a computerized testing center. These tests can include high-quality digital images presented on a computer screen. In other instances, high-stakes tests are completed in a traditional setting where students are presented with a paper test booklet, a bubble sheet, and time limits. Sometimes, students take exams at their own schools with peers. At other times, students go to distant testing centers. High-stakes testing occurs in a wide variety of different settings. It is important to point out, however, that all testing centers and all school classrooms are not the same. Often it may appear as if all students "take" a high-stakes examination in a similar environment. But factors such as room noise, lighting, and temperature may differ greatly from one testing location to another. All classrooms for testing are not equal.

GAINS, LOSSES, AND ASSUMPTIONS FROM AND ABOUT HIGH-STAKES SCIENCE TESTING

Even though high-stakes science testing results are utilized to make important decisions, often the strengths and weaknesses of such data are not fully reviewed. And, as a result, broad conclusions are made that do not reflect the reality of the data. Consider the National Assessment of Educational Progress (NAEP) high-stakes science tests, which are used to, in part, make statewide comparisons. This type of high-stakes test is often used to determine in what manner a school is classified (e.g., underachieving, or meeting minimal state standards). Given that this type of test bears no consequences for individual students, in some instances, students may randomly bubble in answers, spend little time on test items, or sleep through an entire exam. Information garnered from these high-stakes tests may not really reflect true student science achievement. High-stakes science tests also assume that a test taker can read and has completed relevant mathematics classes. Since performance in those two subject areas greatly impacts a student's overall science test score, one must also consider the mathematics and reading level of students. The pressures of accountability occasionally tempt teachers and students to cheat on tests by either obtaining advance copies of the exam or not adhering to the proper test-taking time limits and procedures. Those designing and administering tests make genuine attempts to insure that valid data are collected, but one cannot immediately assume that district scores presented neatly in tables with much technical information are necessarily of high quality.

High-stakes science tests seemingly evolve and change with each administration. The "what" may change from year to year. In some ways this is good, for weaknesses in a test can be addressed. But often such changes in the format and the usage of such high-stakes tests can be counterproductive. Consider for a moment the goal and the reality of high-stakes science teacher certification examinations. Such examinations have been proposed to ensure that "highly qualified" science teachers are hired by school districts. Currently policy makers and others argue this must be done to insure that students can compete at a global level. But often there is such a shortage of science teachers (e.g., physics teachers) that school districts scramble to find anyone who can possibly teach the subject (whether they have passed or not passed the test). In some cases, such certification tests for science teachers do not achieve their goal (to insure the presence of highly qualified science content teachers)—and as a result, the test becomes less high stakes. Word gets out among teachers that the need for any teacher in a particular subject is so great that there is always a loophole of some sort. Changes in the format of a high-stakes science test can greatly impact what takes place in classrooms throughout the year.

For instance, a high-stakes science test might require a "group of students" to demonstrate skills and solve problems together. One goal of such a high-stakes science test structure is to encourage teachers to conduct the type of group lab activities likely to be included on such a test. We believe that was a good goal, for the reality of lab work is that in the real world, employees work in teams. Although the science test might be structured in this manner, what happens if the test is ultimately changed so that individual student scores could be computed and there no longer is "group work"? Teachers will naturally question such changes in test format. They will wonder, "How soon will the test again change? How much effort should I invest in preparing my students for what might be an 'old' test and an 'old' format?" High-stakes science tests can influence what is taught in classrooms, but when a test is altered from year to year (or even cancelled as the result of budget issues), there may be little influence on the science teaching that takes place in classrooms. In some situations, changes in a test could lessen the amount of science that is taught.

Students, teachers, and schools are greatly impacted by decisions made by policy makers utilizing high-stakes tests. Students certainly feel the pressure associated with some high-stakes science exams, including graduation qualifying exams, placement exams, and college admission exams. Students who struggle with standardized testing formats may become less motivated as they experience and anticipate failure. These students may drop out of school when they sense that passing a graduation exam is out of reach. Students may also only learn tested material. Given that many high-stakes tests emphasize a low-level science understanding, these students may not achieve higher order understandings of science content.

Science teachers are also greatly impacted by high-stakes testing. Teachers report altering their teaching practices to better prepare students for exams. These teachers may provide test preparation materials, practice test-taking skills, emphasize topics known to be on the test, and exclude topics that are not tested. Teachers may also teach in ways that violate their own teaching philosophies as they target low-level science understandings for their students (Lomax et al. 1995). As with students, high-stakes testing can be stressful for teachers. Individual schools must react to high-stakes science testing results. Remediation courses are often offered for students who did not pass the exams, and these courses require financial and personnel resources. Schools must also focus schoolwide professional development on areas of improvement for these high-stakes tests.

One concern about high-stakes science testing is that they are not fair across different student groups. Members of different ethnic groups have been shown to approach tests very differently (Boone 1998). Additionally, high-stakes tests are sometimes not satisfactorily modified for English as a Second Language (ESL) learners or for students of special needs. Also, urban, suburban, and rural schools may differ with respect to the amount and kinds of resources that they can use for test preparation and test remediation.

Furthermore, school and parent resources for test preparation vary according to socioeconomic status as well. Clearly, equity is an important issue for high-stakes science testing.

NO CHILD LEFT BEHIND (NCLB): HIGH-STAKES TESTING IN ACTION

The No Child Left Behind Act (NCLB) serves as a particularly timely example of the far-reaching impacts of high-stakes science testing. NCLB is the 2001 renewal of the Elementary and Secondary Education Act of 1965. NCLB lays the framework for school curricula and accountability for all subject areas. NCLB recognizes the importance of science education for elementary and secondary students. NCLB requires mandatory science testing by the 2007–2008 school year for three different age groups: grades 3–5, grades 6–8, and grades 10–12 (U.S. Department of Education 2004). The results of these science tests are used to determine if schools have made "adequate yearly progress," according to NCLB. The adequate yearly progress distinction then influences funding, school management, and even school closure. These high-stakes science tests must be aligned with state science standards. As a result of NCLB, many states are modifying their science standards and putting into place whole new testing systems. Many states have only tested students in selected grades, usually only in math and verbal achievement. As states prepare for the first rounds of state science testing, teachers seek out textbooks aligned with their state standards (and thus the tests). Additionally, many schools require teachers to document how they are addressing state standards in their various science curricula. Thus, teachers are to document how they are preparing their students for the state standards-based, high-stakes science test.

Given the importance of the role of high-stakes science testing required of states by NCLB, one must consider how such tests are commonly developed. It seems, at least in the United States, that most high-stakes science tests are developed by a large testing company such as CTB/McGraw-Hill, tests are developed by a specific state, or a state contracts with a testing company to (in tandem) develop a test that is scored by the testing company. However, as there are many different statewide tests (because state standards differ, and data are collected at different grade levels), it often is difficult to compare high-stakes tests from state to state. To best understand this issue, consider how difficult, if not impossible, it would be to compare one state's third grade fall test results to another state's fourth grade spring test results. The third grade test is administered in the fall to third graders and took two hours to administer; in all, eighty multiple-choice items were completed by students. The fourth grade test administered in the spring consisted of fifty multiple-choice items and two essay items, and students were allotted 1 ½ hours to take the test. No identical items were presented in the two tests. Statewide tests have allowed all of the districts and schools within a state to be compared to some degree. Because state standards frameworks and their accompanying high-stakes tests vary, comparisons from state to state are not always easily done in a manner that is correct.

Although the high-stakes science testing called for by NCLB makes use of reform documents, high-stakes testing does not necessarily target the goals of science education reforms. This disconnect results from both the limited scope of state science standards and the types of tests employed. First, state science standards usually only describe science content that students should know. Many of the other goals of recent science education reform documents such as the *National Science Education Standards* (National Research Council 1996) go untested (for instance, the need for students to conduct inquiry and to integrate math and technology into science). A second reason for the disconnect between high-stakes science testing and science education reform goals is the types of tests used. Often high-stakes tests only include low-level (factual recall) items that do not really get at deep understandings of science material. Why do high-stakes tests, at least in the United States, often only emphasize a low level of science competency? Part of the issue is a pragmatic one: state officials wish to be able to say that a high percentage of their students have passed the test. Also, it is easier (and cheaper) to develop, administer, and score tests that test a low level of science comprehension. Test items that do a very good job of evaluating the sophistication of a student's thinking may be costly, for such test questions may require special equipment and scoring may take time.

WHAT NEXT?

High-stakes science tests provide certain benefits. First, the emphasis placed on high-stakes tests may help ensure that science instruction will receive attention and resources in classrooms. For instance, elementary teachers who may be reticent about teaching science may be strongly encouraged by principals to improve their science teaching. Accountability may also encourage hesitant secondary science teachers to teach controversial topics such as evolution. The emphasis on high-stakes science testing has also focused attention on the standards themselves. Individual states are reevaluating and revising their standards to make them more rigorous and more user-friendly. High-stakes tests themselves are an attractive tool for accountability because they sometimes are relatively inexpensive, are easy to score, and allow standardization across locations. Additionally, the results of high-stakes testing may highlight important gender, ethnicity, or language-learner trends present in science classrooms. Finally, information gained from high-stakes science tests may be used to guide decisions of individual students, teachers, schools, and school systems.

High-stakes science tests also have costs. The financial and personnel costs of implementing the tests, preparing for the tests, and remediation can all be a sizable burden for the limited budgets of departments of education and individual school corporations and schools. Furthermore, increased attention to certain parts of the science curriculum may mean that other science topics are deemphasized or excluded altogether. When high-stakes tests focus on low-level skills, student abilities to do science inquiry and think critically may suffer. Another drawback to high-stakes science testing is the possibility that very important decisions will be made on the basis of the results of a single exam. Furthermore, the emphasis on test scores may undermine the professionalism of teachers and school administrators. Finally, high-stakes testing may not be equitable across different demographic groups.

High-stakes science tests may appear to be stand-alone tests. But in school settings, the high-stakes science test should be considered along with many other issues. We have suggested a variety of whats, hows, and wheres. One particular contextual issue has to do with what other high-stakes tests are administered and when those tests are administered. For instance, a particular state may require the yearly administration of mathematics, reading/language arts, and science tests in grades 3 and 6. In this case, teachers and administrators may prioritize which subjects are to be given more emphasis in the classroom. The authors of this chapter have had many candid conversations with teachers and administrators who indicate that if reading, mathematics, and science tests are to be administered for the same grade level, science preparation will take a back seat to reading and mathematics test preparation. However, it seems to us that at least having high-stakes science testing at the elementary level is better than not having testing at all. If one assumes that a high-stakes science test for the elementary level is not terrible, such a test may increase the low level of attention paid to science by most elementary teachers.

High-stakes science testing has been present in varied forms for many years. With the increase in standards-based assessment taking place at the state level, high-stakes science tests are gaining importance. Such tests will most likely impact what science curriculum is taught in classrooms and the nature of the science teaching that occurs. Of particular importance is the careful review of high-stakes testing data. There are many issues that impact how a student scores on a test and how these results are used.

REFERENCES

Boone, William J. 1998. Assumptions, Cautions, and Solutions in the Use of Omitted Test Data to Evaluate the Achievement of Underrepresented Groups in Science-Implications for Long-Term Evaluation. *Journal of Women and Minorities in Science and Engineering* 4:183–94.

Lomax, Richard G., Mary M. West, Maryellen C. Harmon, Katherine A. Viator, and George F. Madaus. 1995. The Impact of Standardized Testing on Minority Students. *Journal of Negro Education* 64:171–85.

National Research Council. 1996. *National Science Education Standards.* Washington, DC: National Academy Press.

U.S. Department of Education, Office of the Deputy Secretary. 2004. *No Child Left Behind: A Toolkit for Teachers.* Washington, DC: U.S. Department of Education.

ADDITIONAL RESOURCES

American Educational Research Association, American Psychological Association, and National Council on Measurement in Education. 1999. *Standards for Educational and Psychological Testing.* Washington, DC: American Educational Research Association.

Wheeler, Patricia, and Geneva D. Haertel. 1993. *Resource Handbook on Performance Assessment and Measurement: A Tool for Students, Practitioners, and Policymakers.* Berkeley, CA: Owl Press.

3

Conceptual Change as a Viable Approach to Understanding Student Learning in Science

David F. Treagust

DEVELOPMENT OF THE NOTION OF CONCEPTUAL CHANGE

Research on students' and teachers' conceptions and their roles in teaching and learning science has become one of the most important research domains in science education. For more than three decades, researchers have investigated students' preinstructional conceptions on various science content domains such as the electric circuit, force, energy, combustion, and evolution. This research has shown that children are not simply passive learners but make sense of new information in terms of their previous ideas and experiences, which often leads to intuitive knowledge that has been called *children's science*. Indeed, from many studies, science education researchers have shown that students attend science classes with preinstructional knowledge or beliefs about the phenomena and concepts to be taught, that these are deeply held, and that these are frequently not in harmony with science views.

Various theoretical frameworks support research on students' conceptions and conceptual change. Researchers have interpreted students' ideas in terms of the work of Jean Piaget, whereby children are considered to grow intellectually through stages such as the concrete operational stage, when they are dependent on materials for learning, or, at a higher level, the formal operational stage, when they are able to learn without materials simply by thinking about the ideas. Consequently, a key feature in investigating student learning about scientific concepts is the clinical interview, usually on a one-on-one basis when the researcher asks the interviewee questions about the topic of interest. These interviews are referred to as *interviews-about-instances* or *interviews-about-events*. An example of an interview-about-instances is to present a series of cards with sketches about a scientific concept showing various activities that may involve floating and sinking or force.

However, one criticism of this research was that it dealt with student thinking independent of the environment in which the student was engaged. In the past decade, various cognitive approaches merged to take into consideration the social environment within which students learn and to look at both cognitive (Duit and Treagust 2003) and affective aspects (Pintrich, Marx, and Boyle 1993). Recent studies on conceptual change emphasize the importance of the role of the learner, suggesting that the learner can play an active intentional role in the process of knowledge restructuring (Sinatra and Pintrich 2003).

THE CONCEPT OF CONCEPTUAL CHANGE

Research on the concept of conceptual change has developed a unique vocabulary because conceptual change can happen at a number of levels and because different authors use alternative terms to describe similar learning. The most common analysis is that there are two types of conceptual change, variously called *weak knowledge restructuring, assimilation,* or *conceptual capture,* and *strong/radical knowledge restructuring, accommodation,* or *conceptual exchange.*

Consequently, because the term *conceptual change* has been given various meanings in the literature, the term change often has been misunderstood as being an exchange of preinstructional conceptions for the science concepts. In this chapter, the term *conceptual change* is used for learning in such domains where the preinstructional conceptual structures of the learners are fundamentally restructured in order to allow understanding of the intended knowledge, the science concepts under consideration. In a general sense, conceptual change denotes learning pathways from students' preinstructional conceptions to the science concepts to be learned.

EPISTEMOLOGY AND CONCEPTUAL CHANGE

The best-known conceptual change model in science education, based on students' epistemologies—examining how students think about their world—originated with George Posner, Kenneth Strike, Peter Hewson, and William Gertzog (1982). In this conceptual change model, student dissatisfaction with a prior conception was believed to initiate dramatic or revolutionary conceptual change and was embedded in constructivist epistemological views with an emphasis on the individual's conceptions and his or her conceptual development. If the learner was dissatisfied with his or her prior conception and an available replacement conception was intelligible, plausible, and/or fruitful, accommodation of the new conception may follow. An intelligible conception is sensible if it is noncontradictory and its meaning is understood by the student; *plausible* means that in addition to the student knowing what the conception means, he or she finds the conception believable; and the conception is fruitful if it helps the learner solve other problems or suggests new research directions. The extent to which the conception meets these three conditions is termed the *status of a learner's conception*. Resultant conceptual changes may be permanent, temporary, or too tenuous to detect.

Conceptual changes do not occur without changes in a learner's conceptual status. When a competing conception does not generate dissatisfaction, the new conception may be assimilated alongside the old. When dissatisfaction between competing conceptions reveals their incompatibility, two things may happen. If the new conception achieves higher status than the prior conception, conceptual exchange may occur. If the old conception retains higher status, conceptual exchange will not proceed for the time being. It should be remembered that a replaced conception is not forgotten and the learner may wholly or partly reinstate it at a later date because the learner, not the teacher, makes the decisions about the status of the new concept and any conceptual changes.

ONTOLOGY AND CONCEPTUAL CHANGE

Although George Posner and his colleagues (1982) interpret conceptual changes in terms of students' epistemologies, they do include changes in students' ontologies, the ways they view reality. Other researchers, however, use specific ontological terms to explain changes to the way students conceptualize science entities; for example, Michelene Chi, James Slotta, and Nicholas De Leeuw (1994) called their strongest ontological changes *tree swapping*. Two examples for these types of ontological change are that the conception of heat needs to change from a flowing fluid to kinetic energy in transit and a gene from an inherited object to a biochemical process. There are many other concepts where scientists' process views are incommensurable with students' material conceptions, and the desired changes to students' ontologies are not often achieved in school science.

EXAMPLE OF A STUDY OF CONCEPTUAL CHANGE IN OPTICS

One area of research in science education that we have found conducive to engendering conceptual change is with analogies—those situations where the features of a known situation are compared with those of a new situation. Analogies are used in everyday life, as for example when we compare a camera with an eye, drawing attention to the similarities between the lens in each and between the eye's retina and the camera's film. Working with an experienced teacher who was teaching tenth-grade students optics, David Treagust, Allan Harrison, Grady Venville, and Zoubeida Dagher (1996) set out to assess the efficacy of using an analogy for refraction of light that comprises a small cart on wheels painted with luminous paint moving at an angle from a smooth surface, the laboratory bench top, to a rough surface, a piece of carpet. The analogy is that the small cart changes direction when it meets a different surface, as does light when it enters a transparent medium of different density, such as glass, from air. We were interested to learn whether or not the use of the analogy engendered conceptual change in students' learning about the refraction of light. Subsequently, one teacher taught two classes of students over one ten-week term; one class was taught analogically, and one was not. We interviewed all students three months after instruction using an interview-about-instances protocol.

Factors related to status were identified from the interview transcripts to help in the process of classifying each student's conception of refraction as being intelligible, plausible, or fruitful. For example, for a concept to be intelligible, students must know what the concept means and should be able to describe it in their own words; for a concept to be plausible, the concept must first be intelligible, students must believe that this is how the world actually is, and it must fit in with other ideas or concepts that students know about or believe. Finally, for a concept to be fruitful, it must first be intelligible and plausible, and should be seen as something useful to solve problems or a better way of explaining things.

Examples of factors relating to these descriptors from the interviews included the students' ability to draw and describe refraction, comments they made about their own conceptions, and how their language demonstrated conviction of these conceptions. For example, a student whose response to a question in the ray-tracing tasks was that "the beam of light is going through different densities ... so it makes the light bend," but who could offer no further explanation, was given the status of intelligible but not plausible. This student demonstrated that the notion of refraction being related to light changing speed in media with different densities was intelligible to her. However, statements like "I'm not sure" and "It might or might not be" indicated that she did not necessarily believe that this is how the world actually is, and the concept was probably not plausible to the student.

Another student, whose conception of refraction was classified as being intelligible and plausible but not fruitful, talked about her conception of refraction in relation to the wheels analogy. To this student, the idea of refraction was plausible because it fitted in with other ideas or concepts she knew about, that is, the toy car. Statements like "I've seen it before" and "So that's why it is possible" indicated that she believed this is how the world actually is and that it fitted her picture of the world.

Excerpts such as the ones described above were not used in isolation to classify the status of students' conceptions of refraction. Rather, the transcript as a whole was considered by two or more raters and a series of statements, like the ones above, along with a rating sheet, and student drawings were used to guide the rater to classify the status of the student's conception.

Following in these individual interviews using ray-tracing tasks, students who were taught with the analogy demonstrated a higher status for their conception of refraction than those who were not taught with the analogy; 36 percent of students from the analogy group used their understanding of refraction in a fruitful, problem-solving manner, whereas no student from the nonanalogy class could do so. Similarly, 64 percent of the students from the analogy class and only one student from the nonanalogy class showed evidence of plausibility in their explanations of refraction.

It should be noted that a comparison of the means and the pattern of responses for the two classes indicated that while the nonanalogy class scored as well as the analogy class on the interview ray-tracing tasks scored as a conventional test, the analogy group produced explanations of refraction that were of a higher status than those in the nonanalogy group. It is asserted that the analogy enabled these higher status explanations to be generated and that the concrete nature of the analogy enabled the students to remember what they had learned about refraction. As students have difficulty articulating the refraction phenomenon, it would seem reasonable that students who possess a familiar analogy will produce the dichotomy seen in the class results.

Most of the evidence from this study indicated that conceptual change is not necessarily an exchange of conceptions but rather an increased use of the kind of conceptions that makes better sense to the student. This research has shown that while increased status of a conception is possible by means of analogical teaching, it does not necessarily lead to different learning outcomes as measured on traditional tests.

EXAMPLE OF A STUDY OF CONCEPTUAL CHANGE IN GENETICS

Peter Hewson and John Lemberger (2000) analyzed the notion of conceptual change at a more sophisticated level by identifying different descriptors that can be used to gauge the degree of intelligibility, plausibility, and fruitfulness involved in any conceptual change. Chi-yan Tsui and David Treagust (2004) used these descriptors in a study that involved four teachers who engaged students in the learning activities of a computer program called *BioLogica* (Concord Consortium 2001; free download is available at http://biologica.concord.org). This genre of educational software known as a *hypermodel* features multiple representations with rich graphics and animation using fictitious dragons as the major context of investigation for learning introductory genetics in high schools. *BioLogica* allows students to manipulate objects of genetics represented at these different levels of organization—DNA, genes, chromosomes, cells, organisms, and pedigrees—and observe the behavior of these virtual objects constrained by the principles of genetics. In this research, we were interested in the ways in which multiple representations are likely to foster the development of deeper understanding of the gene concept. The four participating teachers incorporated BioLogica in their regular teaching programs in years 10, 11, and 12.

In order to assess the status of students' conceptions at the beginning and end of the instruction in genetics, we sought to identify the status of their conceptions using descriptors that are more informative than those in the optics study. For example, to assess evidence of intelligibility, we looked for evidence of the following four descriptors: representational modes in students' responses

or explanations, analogy or metaphor to represent a conception, pictures or diagrams to represent a conception, real-world exemplars of a conception, or
linguistic or symbolic representations of a conception.

Similarly, to assess evidence of plausibility, we looked for evidence of the
following six consistency factors: reasoned consistency with other high-status
knowledge, consistency with laboratory data or observations, particular past
events from experience consistent with conception, consistency with epistemological commitments, ontological status of objects or beliefs, and causal
mechanism invoked. For fruitfulness, we examined whether or not the power
of the conception has wide applicability, looked to what new conceptions
might do, explicitly compared two competing conceptions, and associated
new conception with experts.

In the interviews following the lessons involving *BioLogica*, students displayed a high status for intelligibility by analogically referring to the dragons
when describing aspects of human genetics. Similarly, several students drew a
diagram to show the relationship between DNA, a gene, and a chromosome.
When a body cell divided, they visualized the double helix of the DNA mole-
cule as a particle in the gene, which was within the wriggling shape of a pair of
replicating chromosomes.

In everyday life, we express our understanding of events in conversation by
referring to examples or exemplars, and this is exactly what we did in the
research when students were asked about their conceptions of genes. Interviewees gave common human hereditary characteristics such as hair and eye
color as real-world exemplars to illustrate their gene conceptions. In this
study, interviewees used language (linguistic or symbolic representations) to
represent gene conceptions, although ontologically they focused on the gene
being a thing rather than a process. For example, one student, when asked
whether BioLogica activities helped his understanding of the gene, used terminology to describe what he had learned: "Well I guess yeah, it [*BioLogica*]
showed us the effects of mixing genes from different parents, and we learned
about the way dominant and recessive genes work, and the difference between
autosomal and sex linked [inheritance]."

In a similar manner, plausible or partly plausible conceptions were evident
when students' conceptions or their thinking about their conceptions could be
mapped to one or more of the six status elements. For example, one student
referred to the consistency of his learning with the observations made during
the computer-based activities when he said, "It wouldn't have sunk into my
head as much as if we just sat in class talking about it, rather than actually
doing it on the computers, because you can actually see from the results of
like a baby Dragon, the effects."

While intelligibility and plausibility of a conception are likened to rep-
resentability and reality/truth, fruitfulness confers the highest intellectual
power of a conception to a person holding it. Only when a conception is
intelligible and plausible can it become fruitful. In a postinstructional interview, one student showed the wide applicability of his conceptions when he

stated confidently that genetics is useful because "it helped me to understand how children get certain things from their parents." When reading a newspaper clipping about genes, another student said, "Ah. We understand much more about this, because they're talking about genes and you know what genes are now." Yet another student commented, "Well, if say I was about to have a child or something, I'd know the precautions to take or, I'd know that there were options. So, it's really useful. I think everyone should be taught about all those opinions in genetics. Should be mandatory."

With regards to fruitfulness of their conceptions, in the postinstructional interview, one student verbalized his vision of studying genetics in the future, its power—"It made me consider the career path of a genetic science as well"—and also said, "My opinions have changed because I now know what's involved with genetically modified food and stuff like that."

CONCLUDING COMMENTS

As would be expected to address the complex phenomena of teaching and learning science, the state of theory building on conceptual change has become more sophisticated, and the teaching and learning strategies developed have become more complex. However, the gap between what is necessary from the researchers' perspectives and what may be set into practice by regular classroom teachers has increased more also. Research has shown that although many conceptual change strategies have been developed and evaluated in actual classrooms, often in close cooperation with teachers and researchers, what works in special arrangements does not necessarily work in everyday practice.

Although conceptual change approaches usually are more successful than traditional approaches in guiding students to the science concepts, there is a problem with research on conceptual change in that it is difficult to compare the success of conceptual change and other approaches. Usually different approaches to teaching and learning address different aims, and hence it is only possible to evaluate whether the particular aims that were set have been adequately met.

Conceptual change primarily has denoted changes of science concepts and principles, that is, cognitive development on the science content level. Often it has been overlooked that these changes usually are closely linked to changes of views of the underlying concepts and principles of the nature of science. The research has not been taken into consideration that understanding science includes knowledge of science concepts and principles and *about* this science content knowledge.

REFERENCES

Chi, Michelene T. H., James D. Slotta, and Nicholas De Leeuw. 1994. From Things to Process: A Theory of Conceptual Change for Learning Science Concepts. *Learning and Instruction* 4:27–43.

Concord Consortium. 2001. BioLogica. http://biologica.concord.org.

Duit, Reinders, and David F. Treagust. 2003. Conceptual Change: A Powerful Framework for Improving Science Teaching and Learning. *International Journal of Science Education* 25:671–88.

Hewson, Peter W., and John Lemberger. 2000. Status as the Hallmark of Conceptual Change. In *Improving Science Education*, edited by Robin Millar, John Leach, and Jonathan Osborne, 110–25. Buckingham, UK: Open University Press.

Pintrich, Paul R., Ronald W. Marx, and Robert A. Boyle. 1993. Beyond Cold Conceptual Change: The Role of Motivational Beliefs and Classroom Contextual Factors in the Process of Conceptual Change. *Review of Educational Research* 63:167–99.

Posner, George J., Kenneth A. Strike, Peter W. Hewson, and William A. Gertzog. 1982. Accommodation of a Scientific Conception: Toward a Theory of Conceptual Change. *Science Education* 66:211–27.

Sinatra, Gale M., and Paul R. Pintrich. 2003. *Intentional Conceptual Change*. Mahwah, NJ: Erlbaum.

Treagust, David F., Allan G. Harrison, Grady J. Venville, and Zoubeida Dagher. 1996. Using an Analogical Teaching Approach to Engender Conceptual Change. *International Journal of Science Education* 18:213–29.

Tsui, Chi-yan, and David F. Treagust. 2004. *Learning Genetics with Multiple Representations: Cross-Case Analyses of Students' Conceptual Status*. Paper presented at the annual meeting of the National Association for Research in Science Teaching (NARST), Vancouver, Canada, April.

4

Finding Meaning in Science

Richard H. Kozoll and Margery D. Osborne

What do we describe when we talk about finding meaning in science, about going beyond knowing facts and definitions, to knowing what something represents or feels like, to having it shape and color our perceptions and understandings? How do we talk about going one step further in education, to talking about how the things we learn in a discipline shape what we do and think, and want to do and think? Furthermore, to go this one step further, to make science meaningful—so that it adds significance or purpose to one's life—how can science do that? These are fundamental questions of education in general, for in what ways do learning and knowing shape and change a life? Such a sense of meaning suggests thinking of science as a way of knowing that enables individuals to enter into the social and intellectual life of their community in a new and different way. It carries a sense of reconstruction of self through a newly articulated engagement with the world and society. Finding meaning in science is a move from procedural knowing to connected knowing through engagement with the discipline. It implies both a deep and reflective understanding and an engagement in a quest, for finding meaning is a quest in the sense that it is never truly fulfilled.

Within science education, the question of how learners find meaning or a sense of value in science so that they choose to engage in science studies would seem to be answered by current concepts of "best practice" in the field. For example, constructivist and situated learning theories hold that establishing connections between science concepts and student's lives promotes personally meaningful, worthwhile understandings. Arguments around the importance of scientific literacy and authentic, project-based learning suggest the usefulness of scientific knowledge in personal and social decision making. Given current science education practices, why is it so hard, then, to develop students' investment in science so that they resonate to our convictions about the importance of science understanding? Using the hermeneutic/phenomenological sense of *lifeworld* as our "being in the world" (Heidegger 1962), we explore questions of meaning in the teaching and learning of science. When we consider such lifeworld experience amid our enacted/socially constructed roles, we suggest that by taking the notion of identity in science to include students' identities in their collective, inclusive of an orientation toward both who the student is and who he or she wants to become, we can enable this broader educative process.

Considering students' everyday experiences and the possible connections to science they might construct, we need to consider that children become situated within a system of values and experiences that define ways of acting. In other words, lifeworlds guide interpretations for both the individual and shared social world, thereby interconnecting these experiences and values that allow our identities, personalities, and social roles to materialize from within lifeworld knowledge (Habermas 1984, 1987). Identity, in practice, is a way of being in the world, a layering of events and interpretations that inform one another and are produced from our participation in communal practices of lived experience (Wenger 1998). Consequently, when we consider issues of meaning in science education, we need to discuss science, its learning and teaching, as a piece of this lifeworld, one that is manifested through expressions of identity that embody a student's understanding of self in relation to others.

Thus, the contexts a student brings to any event of understanding that may involve science include not only their cognitive insights but also the affective tendencies and responsiveness that constitute their impression of a personal identity (Hogan 1998). Children's developing identity in relation to science is set in culturally normative and changing views of self, which include inherent values and insights about what it means for them to be a person. If we examine children's identity as students relative to subject matter, there is a development of a broader understanding that concerns not only the topic of science but also a type of self-understanding involving practical involvement. In this way, finding meaning in science becomes tied to a child's sense of self. For science to act as an intermediary of this type, it becomes a part of the relationship between the student, other people, and the natural world. This whole extends our concerns associated with identity to include the social, humanistic, aesthetic, and spiritual elements constituting a subjectively personal relationship between a passionate, dependent, and intuitive self and others comprising one's worldview (Cobern 1993). This is what might enable engagement with science to act as a medium of deeper personal development that defines authentic engagement, infusing understanding with meaning.

Where do we begin to locate science in relation to the self? To do so, we must begin with the broader discussion of what exactly constitutes the self.

Here we rely on Fay's (1996) articulation of the two distinct manners with which matters of the self may be divided. First, we can speak of atomism and the self as a single isolated entity that, as the subject of consciousness, persists through time and is the source of all our activity. Individuals are distinct and only externally related to others, something to be confronted in an encounter and dealt with, whereas the self is the underlying entity of these conscious confrontations, with any changes of state composed of alterations to this underlying substance. The self is located in the substantial unity of this hard, closed-off, and integrated entity. But substantial unity of this type does not explain how the individual can be composed of multiple identities and the fact that we can be different in different settings, circumstances, and dialogues. How, then, may we go from identity to self in science?

We can answer this question when we shift to the second manner in which self may be envisioned. Here the self is porous, dialectically interacting with others, and relationally rather than substantially unified. The self is created in the very process of interacting with others and the environment. Taking into account the individual standing in relation to certain others, there is no need for an underlying thing, a true self, required for unity. Instead, the self comprises certain states of proper relations. Suggesting this alternative view of self accounts for both a unity of self and the way we are fragmented and fluid. Ultimately, from this perspective the self can be conceived as an on-going activity dedicated to self-creation where others are not separate from but a necessary part of us. So, quite clearly, others have an integral role in this process. Without an intimate connection between self and other, authentic experience could not exist, and so it is here that we can move from identity to self and hence to meaning in the practicing of science. More than pure self-understanding and self-interest, identity is, in part, to be related to others in particular ways and to understand ourselves as related. In these under-standings, we become aware of ourselves as objects to others, thereby become aware of ourselves as selves, and acknowledge the beings of self and other as interrelated to the extent that they affect, and give meaning to, everything about us, including our aesthetic and emotional views.

If we no longer regard self as a unitary phenomenon or structural entity composed of factors and traits that add up to the total person, meaning is fractured. It is created as a social construction through interactions and transactions between and among social beings. Our more autonomous self-interests intermingled within these relationships strike a balance between the individual and others. In this case, meaning is grounded in a dialectical understanding of the object through the construction of self within the social sphere. This view of meaning-in-relationship stresses our essential connec-tions to others, recognizes interdependence in the form of relationships, and acknowledges continuity and difference embedded in situated relationships (Plumwood 1998).

For example, a Mexican American migrant worker student, Andrea, who we have written about (Kozoll and Osborne 2004) talked with us concerning how lonely the migrant life caused her to feel. Andrea wanted her classes, school, and education to do more than teach her facts, theories, and procedures. She also wanted them to provide spaces where she and her classmates could learn from each other, would be colleagues, and in so doing would come to know each other amid relationships based on who Andrea was and thought she might like to become. And in science, Andrea found a class that provided just that.

> Oh yes! I love science, anything that has to do with nature. Science is always a lot of group work, a lot of coming together and getting results from other people you know? "What did you see? Well I observed this. What did you observe, you know?" A lot of interaction and I like that. Through science I got to know a few people.

Here we can find science's meaning, its power, significance, relevance, and authenticity to a student. Science's meaning lies in its ability to become a part of the "ongoing project of constituting and reconstituting selves that are more developed, less alienated, more ethical and humane, less oppressed and exploited, more sensitive and conscious, more free, more creative, and more in harmony with nature and other human beings" (Allen 1997, 22). We can see how this might look in Andrea's sentiments. Her connection to science does not lie in a certain scientific discipline or its topics. In her statement, nothing of the sort is mentioned. Her "love" of science lies first in the connection she has found between science and nature, which is something she has a love for in its own right. Second, Andrea's pleasure in science also lies in science's ability to connect her with other people. It was one way in which she and others came together, conversed in the sharing of ideas, and interacted in such a way that Andrea overcame some of her feelings of loneliness and isolation. In this process of interaction, sharing, and coming to know others and herself, Andrea found meaning in science though these connections.

As another example of finding meaning in science amid self and other, we would like to describe Keith's involvement. Keith is a Jamaican American immigrant who majored in science in college and then went on to become a preservice teacher in our courses. Keith's story exemplifies how engagement with science can become infused into the lifeworld in such a way to become part of his articulations of self.

In his conversations with us, Keith spoke at length about his experiences of nature and the natural world in Jamaica, and how those experiences infused and enriched the science he was learning in his college classes and caused him to enjoy them.

> I like being able to explain common things. I like it when the classes are relevant. I mean I'm not talking about going down to what's in a cell, just why things happen the way they do. I really enjoy that, and in [a] very, very simple way.

The things that Keith found so interesting in his science classes connected to the "science" he enacted elsewhere. Such a science included the ants he observed with his cousins in Jamaica, and further embodied his relationships to other people.

> There's a bench that we always sat on. But there's a tree right next to it and there's these ants that just chew up the wood and their lair is, I don't know, a tunnel. It's just basically a tunnel. They'd build a tunnel and you could see the tunnel going along the tree. Going up the tree or down a branch. And it's made ... I guess what they did is they chewed up some wood and they process it into this little paper like material and just build that whole tunnel. And we'd go out there and mess with those, which would be stupid 'cause they'd bite you, but we'd go out there and we'd mess with those things.

His science classes connected with such experiences and took them further, explaining things he saw and provided him with a deeper understanding.

> And it's funny, I just found this out too in my entomology class, I just found out that once somebody breaks into [the ant's] house, they send out these pheromones and they'd call the army ants, the soldier ants, or whatever. When I heard that, I was thinking back, I mean that was what would happen ... you'd break a hole and just instantly a whole bunch of these big ants would come out and they'd come through that hole and, you know, try to see who was there. I mean I got a bunch of bites from them!

Keith called this "confidence," and he claimed that his classes gave him confidence in his knowledge of the things he observed by giving him new lenses to understand those things.

> Now I think about the whole thing. I can definitely see how [theories from classes] would change my perceptions on nature. I was thinking about this when I was talking about confidence.... It's just because I've started to just notice things a lot more and I've started to think about things a lot more. I just make connections I've never thought of before.

This in turn gave him different perspectives to theorize his own relationship to nature.

> I mean it all goes back down to just knowledge. You look at things, different things, from different angles.... Then you can look at things from like a completely different perspective but still bring it back to just the natural thing.

Keith used his new lenses on the world to notice and explain things other people would not have cared about, like how the grass is growing in someone else's yard, the population dynamics of trees by the side of the highway, and what happens to hair clippings. He showed great personal satisfaction in his ability to do this and share his ideas with others. Keith has found a meaning in science that is embedded within a relationship that exists between himself, science, and the world around him as well as family, friends, and strangers alike who both have and do not possess such a scientific outlook. We asked him about such a perception:

> I think that's the driving force you know. It's like you have this picture that probably will never be finished and you don't know what it's of. But maybe if you look at the _whole_ picture, you'll see what it's _about_. I mean I just feel that that's been driving me.

Through such a connection, Keith's understanding of science infuses meaning into what he thinks and experiences. It is part of his identity—the lens he has constructed to view the world. Keith's ability to use science in order to bring together disparate contexts in order to make sense of his experiences, and further connect these experiences to other aspects of his life, provides science with meaning and a place in his lifeworld. In this case Keith used his entomology class to understand the ants he observed in Jamaica, and his ecology class to understand the grass growing in someone's yard. Further, when Keith speaks of these connections, he also refers to his own feeling of confidence and caring. When talking about the science he has done in school and home, these feelings are made sensible and accessible against the backdrop of science and turn science into part of both the person Keith is and the person he wants to become (Taylor 1994). As part of his identity, science is one way Keith negotiates a deeper and more harmonious existence with the people and places constituting the world around him. He can look upon the world from different angles and perspectives, and this takes him closer to the whole picture Keith is so driven to complete.

Narratives such as Keith's and Andreas's position science amid holistic concerns, which further reflected critical understandings of their social circumstances. The meaning they found in science included but extended beyond their knowledge of scientific facts, theories, and ideas. Taking into account how these relationships between student and science formed directed our attention to the larger perspectives and frameworks from within which meaning and value in science are constructed. For science to have meaning, it must become a way of understanding our relationship to self and others.

REFERENCES

Allen, Douglas. 1997. Social Constructions of Self: Some Asian, Marxist, and Feminist Critiques of Dominant Western Views of Self. In _Culture and Self: Philosophical Perspectives, East and West_, edited by Douglas Allen, 3–26. Boulder, CO: Westview Press.

Cobern, William W. 1993. College Students' Conceptualizations of Nature: An Interpretive World View Analysis. _Journal of Research in Science Teaching_ 30:935–51.

Fay, Brian. 1996. *Contemporary Philosophy of Social Science*. Oxford: Blackwell.

Habermas, Jürgen. 1984. *Reason and the Rationalization of Society*, vol. 1 of *The Theory of Communicative Action*. Boston: Beacon.

———. 1987. *Lifeworld and Systems: A Critique of Functionalist Reason*, vol. 2 of *The Theory of Communicative Action*. Boston: Beacon.

Heidegger, Martin. 1962. *Being and Time*, translated by John Macquarrie and Edward Robinson. London: Routledge and Kegan Paul.

Hogan, Paul. 1998. The Politics of Identity and the Experience of Learning: Insights for Pluralism from Western Educational History. *Studies in Philosophy of Education* 17:251–59.

Kozoll, Richard H., and Margery D. Osborne. 2004. Finding Meaning in Science: Lifeworld, Identity, and Self. *Science Education* 88:157–81.

Plumwood, Val. 1998. Nature, Self, and Gender: Feminism, Environmental Philosophy, and the Critique of Rationalism. In *Environmental Philosophy: From Animal Rights to Radical Ecology*, edited by Michael E. Zimmerman, J. Baird Callicott, George Sessions, Karen J. Warren, John Clark, and Karen Warren, 291–314. Upper Saddle River, NJ: Prentice Hall.

Taylor, Charles. 1994. The Politics of Recognition. In *Multiculturalism*, edited by Charles Taylor, 25–73. Princeton, NJ: Princeton University Press.

Wenger, Etienne. 1998. *Communities of Practice*. New York: Cambridge University Press.

5

Controversy over Evolution

John R. Staver

ORIGINS

Controversy over evolution is a debate about origins, the origin of humans and life. But, what are the origins of the debate itself? The origins are not documents such as *The Fundamentals*, which were written by evangelical Christians in the early 1900s, not Charles Darwin's 1859 publication *The Origin of Species*, and not the First Amendment of the Constitution of the United States of America, as important as these documents are to the controversy. The origins of the controversy are the fundamental concepts of faith and reason and their interaction.

FAITH, REASON, AND THEIR INTERACTION

Faith is defined in Hebrews 11:1 as follows: "Now faith is being sure of what we hope for and certain of what we do not see" (*Holy Bible*, 1984). Reason is a capacity whereby humans move from a premise to a conclusion by thinking. Visualize a continuum—or perhaps an old-fashioned teeter-totter—of authority or dominance, with faith and reason at opposite ends. Faith and reason have interacted throughout the history of humanity, and a controversy that first exploded nearly a century ago in Dayton, Tennessee is rooted in documents written more than two millennia ago, the Old Testament and the writings of Plato and Aristotle.

Regarding faith, God directs the land to bring forth living creatures according to their kinds, then provides examples of kinds such as livestock, wild animals, and creatures that move along the ground in Genesis 1:24. Probably composed and written during Judah's exile in Babylon (587–538), not only does the first creation story's introduction of the concept of kinds predate the Greeks, but also the Old Testament Scriptures were not translated from Hebrew and Aramaic into Greek until about 250 years prior to the birth of Jesus.

Regarding reason, Plato (428–427 to 348–347) described pure, fixed Forms in an ultimate world beyond human understanding, and Plato employed an allegory of shadows on the wall of a cave to explain how humans encounter this ideal world. The light that produces the shadows represents the ultimate world or Form of the Good, but the shadows themselves represent the indirect or secondhand knowledge of most humans. A few humans are able to emerge, first to see things in the cave or world as they are, then from the cave into the direct light of the sun to see the final source of all reason and truth, which is the Form of the Good. Aristotle (384–322) shared Plato's notion of pure, fixed Forms and applied it to understanding nature's purpose, arguing that understanding nature required understanding why things exist in addition to their forms.

FAITH AND REASON IN MEDIEVAL AND MODERN THOUGHT

Whereas reason held an upper hand in Greek philosophy, the dominant authority of medieval European philosophy prior to the Enlightenment was the perspective from which philosophers reasoned. To a person, they were clerics, with positions in universities or monasteries, and the authority and context within which they reasoned was Christian doctrine. Early Christian thinkers linked Plato's pure, fixed Forms to God's creation in Genesis, including God's plan and God's ideal forms, all of which lie beyond human understanding. They connected Aristotle's description of nature's purpose in terms of its Form to God's purpose, which was to create humans to worship Him. Plato's and Aristotle's notion of constant Forms was linked to the Genesis presentation of kinds of organisms as God's creation of all things living and non-living, despite Aristotle's view that biological classes were not unchanging Forms but were flexible and could contain intermediates.

In contrast, early Enlightenment European philosophy was dominated by a setting aside of Christian doctrine as the authority and perspective within which to reason. Most still believed in God, except for David Hume, who stood alone as an atheist, and the concept of God retained a fundamental place for major philosophers of the time such as René Descartes, Benedict de Spinoza, and George Berkeley, who was a bishop. Descartes, John Locke, Immanuel Kant, and others conducted their studies of nature, called natural philosophy, within their philosophical perspectives. The invention of the printing press also allowed early modern philosophers to communicate the written word to a broader readership. Finally, nearly all of the early modern philosophers were laymen, save Berkeley, and almost none held positions in universities, except Kant.

DAWN OF MODERN SCIENCE

As reason established a dominance over faith during the Enlightenment, reason also developed its own continuum as the rationalism of thinking in philosophy and a new concept, empiricism—the founding of knowledge on sensory experience—began to gain credence as modern science was born in the late 1500s and early 1600s.

DARWIN'S THEORY—SETTING FAITH ASIDE IN BIOLOGY

In the nineteenth century, a major figure in this process was Charles Darwin, the son of a medical doctor, who began medical studies at Edinburgh University, withdrew, and then enrolled at Cambridge to pursue religious studies. But his mind had turned toward the study of nature and after completing his studies at Cambridge in 1831, Darwin embarked shortly after Christmas that year on a surveying voyage to South America sponsored by the British government. Darwin served as the ship's naturalist. When the HMS *Beagle* concluded its voyage in October 1836, Darwin disembarked not only with a wealth of scientific data but also puzzling scientific questions, among them questions about the origin of the plants and animals on the Galapagos Islands. Over the next twenty-three years, Darwin formulated, tested, and discussed his work. Darwin's hypothesis of natural selection stemmed from his realization that the numbers of plants and animals in populations changed only modestly, if at all, over periods of time despite their enormous reproductive potential. Darwin reasoned that modest, if any, change in the face of such enormous potential was because only a small fraction of the members in a population survived, reproduced, and left offspring. He further reasoned that the survivors were those individuals of a population who were more fit. In other words, an environment naturally favored—selected— the survivors based on their fitness to function in that environment. Unable to test directly this hypothesis of natural selection, Darwin tested it indirectly by first assuming that natural selection did operate, then predicting the occurrence of patterns, and then checking to see if such patterns did occur. Through extensive communications with other experts, reading, observations, and experiments, Darwin enlarged his vision of the problem and proposed a theory of evolution by natural selection as an explanation. In 1859, Darwin formally published his evidence, reasoning, results, and theoretical explanation in *The Origin of Species*.

Darwin's work contributed three important elements to the advancement of biological knowledge. First, evolutionary theory not only explained the origin of species but also brought coherency to facts in the biological sciences that were already known in his day. Second, Darwin's theory raised new questions for future research and served as a guiding framework for scientific studies that have enhanced our understanding of the origin of genetic variation, laws of heredity, and rates of evolutionary change. Third, Darwin's theory fundamentally altered the prevailing philosophy of science of his day with respect to origins. Until Darwin, leading naturalists such as Louis Agassiz explained the observed patterns among organisms and the origin of species as part of God's design and action, asserting that God had created all species fitted for certain environments at specific times. Darwin explained these concepts through empirical evidence and reasoning from such evidence, excluding any reference to God's plan or actions.

Over the latter part of the nineteenth century and into the twentieth century, biologists empirically tested Darwin's theoretical explanation and its predictive usefulness, and subsequently adopted his theory of evolution by natural selection and his perspective of explaining biological phenomena without reference to God's design or action. This work was, however, not without scientific criticism. Scientific skepticism concerning the importance of natural selection led to attempts by scientists to deemphasize the role of natural selection. The reconciliation of this question produced a modern synthesis of evolution in the 1930s, which instead documented the importance of and reiterated the central role of natural selection in acting on genetic variation.

CHRISTIAN FUNDAMENTALISM
IN AMERICA—A REACTION TO EVOLUTION

Whereas biologists embraced evolution, some religious communities viewed evolution as another step in modernism's quest to undermine religion. The origin of Christian fundamentalism in the United States in the late 1800s and its growth through the early 1900s reflect the growing antievolution sentiment. American Christian fundamentalism can be traced to an extensive, cross-denominational network that formed and grew around the revivalist crusades of Dwight L. Moody, the founder of the Moody Bible Institute in Chicago and perhaps the era's greatest evangelist. Working largely within a Calvinist tradition, Moody and other early leaders advocated evangelism as the church's top priority, a born-again perspective, the imminent return of Christ, and an inerrant Bible. Despite the movement's growth as the nineteenth century ended and the twentieth century began, its leaders worried about the growing secular psyche and liberal church doctrine in America as well as a weakening commitment to evangelism within their own movement. Their response was *The Fundamentals*, a series of twelve separate booklets written and published between 1910 and 1915 and later edited by Charles Feinberg, retitled *The Fundamentals for Today*, and published in a two-volume set in 1958. *The Fundamentals* criticized these trends and reaffirmed their own beliefs that, (1) Scriptures are inerrant, (2) Jesus is the Son of God, (3) Jesus died on the cross in our place and for our sins, (4) Jesus was resurrected in body and spirit, and (5) Jesus will return. Approximately one-fifth of the essays in *The Fundamentals* criticized evolution as a concept, and this series of booklets is thought to be the origin of the term fundamentalist. *The Moody Handbook of*

Theology dates the first use of the word fundamentalist to 1920 by Curtis Lee Laws, who edited a Baptist newsletter.

The shift from dislike of and disagreement with evolution as a segment of modernist thinking to an active antievolution crusade by Christian fundamentalists in the early twentieth century stemmed largely from the growth of public high schools across the nation and the favorable presentation of evolution in biology textbooks. Large numbers of American students were learning about evolution for the first time, and many of these children were the sons and daughters of parents with fundamentalist beliefs. One thrust of this crusade was extensive discussion by state legislators in over twenty states about banning evolution from the schools. Tennessee, Mississippi, and Arkansas banned evolution, while Oklahoma banned textbooks containing evolution.

CONTROVERSY OVER EVOLUTION GOES TO COURT

The 1925 trial of John Scopes, a Dayton, Tennessee biology teacher, placed this antievolution crusade before national and international audiences. Simultaneously, the Scopes trial shifted public confrontations between supporters and critics of evolution to the courts of the land, a venue that continues to the present day. As a result, much of the history as well as an understanding of the controversy from 1925 to the present can be garnered from the decisions of courts that have heard the arguments and passed judgments on them.

Beginning with Scopes, all of the cases have been argued on First Amendment grounds. The First Amendment states, "Congress shall make no law respecting an establishment of religion, or prohibiting the free exercise thereof; or abridging the freedom of speech, or of the press; or the right of the people peaceably to assemble, and to petition the government for a redress of grievances." The first, second, and third clauses are known as the Establishment, Free Exercise, and Free Speech Clauses, respectively, and all decisions have been based on the courts' interpretations of some or all of these clauses. The combined meaning of these clauses is the concept of religious neutrality, and the sequence of court cases throughout the twentieth century clearly demonstrates a variance of meaning with respect to neutrality in the minds of different people.

One thread of a strong thematic fabric of understanding centers on how antievolution thought and action have changed over time in response to court decisions. Three principal strategies can be identified: (1) prohibit the teaching of evolution; (2) advocate that equal time be given for creation science alongside evolution in science; and (3) avoid all references to religion, declare that scientific controversy exists over evolution, and call for teaching all of the evidence for and against evolution.

The Scopes trial represents the first strategy and the first legal confrontation between representatives of Christian fundamentalism and those of evolution. Its outcome produced a victory for the fundamentalists, despite that fact that the Tennessee Supreme Court overturned Scopes' guilty verdict and $100 fine on a technicality, because the judge, not the jury, imposed the fine. Legal scholars consider the Scopes case unimportant because it lacked a definitive, constitutionally based ruling on the Tennessee law banning the teaching of evolution, but its effect on the treatment of evolution in biology textbooks was important, as evolution was deemphasized over the next twenty-five years. In the 1950s, competition and an increasingly adversarial relationship between the United States and the Soviet Union became known as the Cold War. American fears of a Soviet-dominated world resulted in a new emphasis on preparing U.S. students for college and careers in science, mathematics, and engineering. As a result, evolution regained an emphasis in high school biology led by a then new curriculum development organization called the Biological Sciences Curriculum Study.

Attempts to prohibit evolution continued into the 1960s, and an Arkansas statute fashioned after the Tennessee law prohibiting teachers in public schools and universities to teach Darwin's theory of evolution became the next significant court test. In 1968, in *Epperson v. Arkansas*, Susan Epperson, a biology teacher, sued the State of Arkansas to invalidate the law. Upon appeal, the U.S. Supreme Court invalidated the Arkansas statute because it did not have a secular purpose and therefore did not satisfy the Establishment Clause. The Court also held that the law interfered with teachers' academic freedom to discuss evolution in biology classes and therefore did not satisfy the Free Exercise and Free Speech Clauses.

In 1971 the U.S. Supreme Court established a legal precedent that future courts have applied in rendering decisions about this controversy when it developed and used a three-part argument to interpret the Establishment Clause in *Lemon v. Kurtzman*. The three parts of the so-called Lemon test with respect to a law or practice are as follows: (1) does the law or practice have a secular, not a religious, purpose or intent; (2) is the effect of the law or practice neutral in that it neither promotes nor inhibits religion; and (3) does the law or practice create an undue entanglement between government and religion?

Reacting to the Court's opinion in *Epperson v. Arkansas* and particularly to the discussion of free speech, representatives of antievolution shifted to their second strategy, equal time. The U.S. Supreme Court struck down the concept of equal time in 1987 in *Edwards v. Aguillard*. Applying the Lemon test, the Supreme Court held that a Louisiana equal time statute did not have a secular purpose.

Following the Court's decision in *Edwards v. Aguillard*, antievolution forces shifted to a third strategy, one they are pursuing to the present day. Following this strategy, they avoid all references to religion, declare that scientific controversy exists over evolution, call for teaching all of the evidence for and against evolution, and call for alternative scientific theories such as intelligent design to be taught alongside evolution. The current theme brings forward the concept of equal time but discards all of the details of past creation science.

Intelligent design has become the broad concept under which the extensive variety of antievolution views exists in safe harbor. The history of intelligent design goes back at least to the early 1800s and William Paley, a British natural theologian. Paley presented an argument by analogy for the existence of God from the observation of design. In sum, the argument is as follows: a person walking in a meadow finds a watch on the ground. The watch stands apart from its surroundings due to its complexity. It is too complex to have occurred by natural processes; therefore, it must have a maker. By analogy, the world is too complex to have occurred by natural processes, so it, too must have a maker, who is God. In its current dress, intelligent design is being proposed as a scientific alternative theory to evolution. In summary, the concept of intelligent design is deeply rooted in religion, and thus far it has not been empirically examined and presented to the scientific community by its advocates in ways that would allow the scientific community to evaluate its ability to explain and predict natural phenomena.

INTO THE FUTURE

As this chapter is revised in the spring of 2006, the new federal education law No Child Left Behind (NCLB) and the states of Ohio and Pennsylvania have witnessed attempts to modify science curricula as a result of the current strategy. When the U.S. House and Senate versions of NCLB were reconciled in a conference committee, the report of the conference committee included language stating that the teaching of controversial concepts such as biological evolution should include a full range of existing scientific views as well as the reasons behind a controversy. Whereas the conference committee report is not part of the actual NCLB law and therefore does not carry any force of the law, supporters of antievolution have implied in Ohio, Pennsylvania, and Kansas that it is the law.

Recently, legal and public attention centered on Dover, Pennsylvania because a lawsuit filed against the Dover Area School District became the first court test of the constitutionality of intelligent design. On October 14, 2004 the Dover Area School District Board of Directors approved a resolution that states as follows: "Students will be made aware of gaps/problems in Darwin's theory and of other theories of evolution including, but not limited to, intelligent design. Note: Origins of Life is not taught" (*Kitzmiller et al. v. Dover Area School District et al.* 2005). The Dover school board subsequently placed several copies of *Of Pandas and People*, an intelligent design textbook, in the high school library, and adopted a four-paragraph policy to inform ninth graders in science classes about intelligent design and the availability of *Of Pandas and People* in the library. The district's science teachers refused to read the policy to their students; thus, an administrator read it to all ninth graders. Several parents objected to the board's actions, and the American Civil Liberties Union filed suit against the school district on behalf of the parents. A bench trial was conducted in U.S. District Court in fall 2005, and Judge John E. Jones III issued his 139-page decision on December 20, 2005 (*Kitzmiller et al. v. Dover Area School District et al.*). In summary, Judge Jones ruled as follows: (1) The district's intelligent design policy is unconstitutional according to the Establishment Clause of the First Amendment of the U.S. Constitution and Article I, Section 3 of the Pennsylvania Constitution; (2) intelligent design is not science; and (3) intelligent design is a strong endorsement of religion.

The force of law of this ruling applies only to the Middle Federal District of Pennsylvania. However, Judge Jones' application of specific legal tests to make his decision provides implications for all states and local school districts. His use of the Lemon test and the Endorsement test called into question the school board's purpose for singling out of evolution for special scrutiny and criticism that were not applied to all scientific theories. These legal tests were also employed to examine the school board's purpose for re-defining science to allow supernatural explanations of natural phenomena. According to the Lemon and Endorsement tests, this purpose must not be religious, but Judge Jones found the Dover school board's purpose to be strongly religious in motivation and character. School boards throughout the nation should be aware of these implications.

Boards of education should also understand that prior decisions set precedents that are considered when courts rule on cases at trial. As legal scholars predicted, prior decisions in *Epperson v. Arkansas, McLean v. Arkansas, and Edwards v. Aguillard* weighed heavily in *Kitzmiller v. Dover.* Deciding any future Establishment Clause cases, the federal courts will consider prior decisions that pointed out that concepts of God do not divest themselves of their fundamentally religious character when they are presented as a philosophy or as a science. The core belief that a supernatural creator is responsible for creating humans is a religious belief. Whereas much of the public discussion in response to attempts to insert intelligent design into school science focuses on whether or not it is science, the courts will ask if it is religion.

REFERENCES

Darwin, Charles. 1859. *On the Origin of Species by Natural Selection, or, The Preservation of Favoured Races in the Struggle for Life.* London, England: J. Murray.

Feinberg, Charles L. 1958. *The Fundamentals for Today* (2 vols.). Grand Rapids, MI: Kregel Publications.

The Holy Bible. 1984. New International Version. Grand Rapids, MI: Zondervan Bible Publishers.

Kitzmiller et al. v. Dover Area School District et al. No. 04cv2688 (M.D. PA., December 20, 2005).

ADDITIONAL RESOURCES

Behe, Michael J. 1996. *Darwin's Black Box.* New York: Simon & Schuster.

Bunnin, Nicholas, and Eric Tsui-James, eds. 1996. *The Blackwell Companion to Philosophy.* Oxford, UK: Blackwell Publishers Ltd.

Dembski, William A. 1999. *Intelligent Design: The Bridge between Science & Theology.* Downers Grove, IL: InterVarsity Press.

Forrest, Barbara and Paul R. Gross. 2004. Creationism's Trogan Horse. New York: Oxford University Press.

Mix, Michael, Paul Farber, and Keith King. 1996. *Biology: The Network of Life.* 2nd ed. New York: HarperCollins.

Numbers, Ronald L. 1992. *The Creationists: The Evolution of Scientific Creationism.* Berkeley, CA: University of California Press.

Pennock, Robert T., ed. 2001. *Intelligent Design Creationism and Its Critics: Philosophical, Theological, and Scientific Perspectives.* Cambridge, MA: MIT Press.

6

Learning from Laboratory Activities

Wolff-Michael Roth

Shocked by the launching of *Sputnik* by the Russian space agency during the Cold War era, American science educators started to rethink the way in which science was taught. Until that time, school science consisted mostly of lectures. In the attempt to help students *do* science and thereby prepare more of them to become scientists, engineers, or technicians in the race against the Russians, science educators began to focus on laboratory activities. They hoped that the hands-on work in the laboratory would entice more students into pursuing science-related careers; and they hoped that students would gain a better understanding of science. However, teaching science through laboratory activities never really caught on. Appearing in 1990 and 2004, the journals *School Science and Mathematics* and *Science Education*, respectively, published important articles that reviewed the science education literature on laboratory activities. These articles suggested that the promise of learning science through laboratory activities had never been realized. Very little had changed over the fifteen years between the two reports. Science teachers generally use laboratory activities to motivate students, but do not really believe that they would learn the kind of things they needed to know on tests (Tobin 1990). Students frequently are not serious about their work, believing that the laboratory activities will not improve their test scores. Why, then, should we use laboratory activities at all if they do not appear to help understanding? Or, perhaps we should ask whether there is something wrong about the way laboratory activities are designed and used. Ultimately we might ask, "What needs to be done to get the most from laboratory activities?"

HOW REAL SCIENCE GETS DONE: SCIENTIFIC LABORATORY

There are many myths about science and scientific research, about its objectivity, and about the truth of the facts it discovers. Researchers create these myths by studying the methods that scientists describe in their publications and by asking scientists, once they have become famous for some discovery, about what they have done. Ethnographers who visit scientific laboratories and describe what is going on provide a very different picture of how science gets done (e.g., Knorr-Cetina 1981). They underscore the contingent and contextual nature of science. They show that there is a situational logic at work in the laboratories best described as "making do" rather than a specific rationality. Local idiosyncrasies, know-how, interpretations, and material and social resources determine the accomplishments within scientific laboratories. The myth of scientific rationality and problem solving is established when the contextual factors are stripped and when scientists report their new constructions as if they were the product of unaltered intentions. Let us take a look into one laboratory where, for nearly five years, I participated in all parts of the scientific research and simultaneously conducted ethnographic research.

The biologists I studied were interested in finding out about the changes that salmon undergo during their lifetime, especially around the time these fish go on their long migration to the oceans before returning to their birthplaces. The biologists measured the amount of light that is absorbed by the different photoreceptors in the retina. It turned out that measuring light absorption was not an easy job. The different photoreceptors initially looked little different under the microscope (Figure 6.1, right). To distinguish the photoreceptors, the scientists needed to look at the graph that showed the amount of light absorbed. The amount differs for the different colors in the spectrum. The color where the maximum absorption occurs is supposed to differ for the different photoreceptors (Figure 6.1, left). The scientists, however, were caught in a quandary. To know what the graph displayed, they needed to know what they were seeing under the microscope; but to know what they were seeing, they needed to know what the graphs were displaying. It took several weeks before they learned to make the correspondence between the graphs and the images they got from the microscope. That is, in contrast to school laboratories, where students are expected to learn something within an hour, scientists take weeks, months, and even years to learn something that is relevant to their work.

In part, the biologists' problems arose because some of the patterns on the graph were thought to be due to the instruments, such as fluctuations in the light source. But they were not sure. For example, they bought a new and expensive light bulb; but the patterns in the graph remained. They also thought that the instrument recording the different colors of the light spectrum shifted during the data collection. So they shifted part of the measurement by hand until they got the best-looking graph.

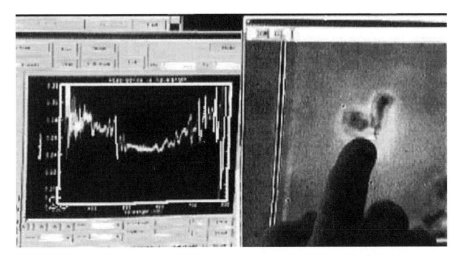

FIGURE 6.1   During an experiment concerning fish eyes, a scientist points to the microscope image of a piece of retina (right). On the left, there is a graphical recording of the amount of light that was absorbed in the piece. To know what kind of piece the scientist is pointing to, she needs to know what the graph shows; but to know what the graph shows, she needs to know what kind of piece she is pointing to.

Sometimes the biologists were completely in the dark about what was going on. For example, one morning as they started collecting data, their computer showed no microscope image—there was "snow" on the monitor where the image of the photoreceptor should have appeared, as it does in Figure 6.1. The scientists fiddled with the different buttons on the microscope and the camera, and worked through the different menus on the computer monitor. They had no idea what had changed since the night before, when everything was still working. For fifteen minutes, they just tried this and that. There was nothing methodical about their work. Then, all of a sudden, everything worked again. The scientists did not know what brought everything back to normal. But because everything worked, they did not worry. They simply began their day's work.

Although the biologists had set out to verify the facts other researchers had reported, their experiment ultimately showed that the research they had attempted to reproduce had been wrong. But because my scientists were hanging on to their initial ideas about what they would find, it took almost two years until they finally came to definite conclusions. In part, they had to become better at experimenting and rule out the effect of the instruments on their graphs. In part, the scientists had to change their ideas and theories about what was going on. In part, they had to become better at their laboratory skills. The ultimate result that they wrote up emerged from an intertwining of all these changes. It was the result of a complicated learning process that involved changes in the instruments, knowledge, and skills. They might have given up, but because they knew what they wanted and were really interested in the research, they were also motivated to continue the research until they had some definite results. They knew there would be a payoff whatever they found.

THE COOKBOOK APPROACH

The traditional and predominant way of teaching science by means of laboratory activities is very different from what scientists do. Students are asked to follow step by step a set of procedures, at the end of which they will have seen something or have collected some data. Based on the observation or the data collected, students are supposed to learn some scientific concept. More often than not, students struggle with getting the activity done. They do not know exactly why they do this experiment. This would require knowing the concept the activity was designed to teach. Students therefore also do not know why they do one step in the procedure rather than another; and they have no way of assessing whether something they have done is what they were supposed to do. For knowing that you have done what you are supposed to do, you need to know that what you see is what you were supposed to see. But to know that you have seen what you were supposed to see, you need to know that you have done what you were supposed to do. Because students do not and cannot know the motive for the investigation, it is not surprising if they are not motivated. Knowing the motive of an activity and motivation and emotion are all tightly interrelated.

The quandary is similar to the one I described in the scientists' case, though it is of a different order—the scientists do not know what they will eventually know, but students are supposed to get the result their teacher expects from them. Readers familiar with trying to follow a new recipe or reading instructions for a new software package, how to assemble some furniture, and so on have found themselves in a similar quandary. Because of the similarity between the step-by-step instructions in laboratories and the instructions in cookbooks, this way of using science laboratories is frequently referred to in a disparaging manner as the *cookbook approach*. Research showed that, at best, students learn how to follow instructions. Teachers, who know what the outcome of the experiments *ought* to be under the best of circumstances, assess students on how close they came to the expected outcomes. Because students are also worried about their grades, it is not uncommon to see them fudge observations and data. This is especially the case in classes where the teacher insists on students getting the "right" results even if there is evidence that the actual results differ under the given conditions. I was once involved in such a situation. Some students who I did not teach were redoing a lab after school while I was preparing the lab activities for the next day. Not knowing what I should be seeing according to the textbook, my observations confirmed those of the students, that is, we saw something different than we were supposed to according to the book. But students are concerned with getting good grades and doing well on tests; learning how science is really done may interfere with this concern. As a result, very little science is learned. Laboratory activities in which teachers and manuals prescribe what students are to do rank very low on a four-point scale that was developed more than thirty years ago (Herron 1971). The lower two scores are given to laboratory activities that provide both problem statement and procedures; students are expected to get a specific result. The upper two scores leave the method of inquiry to the student, but the highest score is reserved for experiments that address problems and questions that students frame themselves. In more recent years, this type of experiments has been characteristic of *authentic school science*.

AUTHENTIC SCHOOL SCIENCE

Some science teachers and science educators endeavor to provide students with laboratory experiences that bear family resemblance with what scientists do, a resemblance that the term *authentic school science* was designed to capture (e.g., Roth 1995). Rather than being given a problem by the teacher or lab manual, students in authentic school science typically design their own experiments to find out about something they are interested in. Because they have a clear motive, students are also motivated and will pursue the investigations over longer periods of time. They know what their question is, so when something does not work in their investigation, they try to figure out what is not working and then redesign their approach. Let us turn for a moment to the students in one of several seventh-grade classes in which I recently taught an integrated science unit (see also Chapter 8, "Scientific Literacy").

In the community where the students and I live, there are problems with the water. In the summer, it rains very little and sometimes not at all. As a result, the creek that drains the watershed within which the community lies is reduced to a trickle. The few resident trout have difficulties surviving in the small number of remaining shaded pools. There is insufficient water in the local aquifer to supply water to the farms. Runoff from the fields, coliform bacterial pollution from chicken and cattle farms, industrial pollution from a small industrial park, and straightening of the creek in many places have led to the sorry water situation.

In a newspaper article, a local environmental group has called all citizens to take actions, in particular to contribute to creating knowledge about a variety of aspects. Unfailingly, students are enticed into contributing when they hear and read about this. They want to contribute, and in the process do a form of science that is actually very useful. They have a clear motive for what they are doing, and it is not surprising, therefore, to see them both motivated and emotionally engaged. Both the precarious situation of the creek flowing through their community and the newspaper article constitute a collective motive, a concern shared by many others in their community.

The students begin generating ideas, often related to cleaning up the creek and to finding out more about Hagan Creek and its problems. After a field trip to different sites along the creek and some library research, they frame initial investigations that then lead into programs of research. As their teacher, I assisted students in reworking their questions to make them productive. This development of a cluster of questions is evident from an initial note that one student wrote:

> We are studying Hagan Creek to find out about what water and creatures are like at the different sites. One of the things we are trying to find out is the quality of the water. The water quality determines what creatures live there. The quality depends on the depth, the width, the bottom (whether it is sandy, rocky, or gravelly), the temperature and the speed of the current. We will take samples of the creatures and then the next day count them and look at them under a microscope. We will make graphs displaying all the different information we got. (Magda, May 5, 1998)

To further assist students in posing interesting questions, I provided two boxes of resources. One box contained a variety of tools including measuring tape, ruler, stopwatches, Serber samplers (a square sampling grid to which a net is attached at a right angle so that specimens from the grid, when stirred up, will float into the net), dip nets, string, pH meters, a pH paper kit, moisture meters, a dissolved-oxygen meter, a colorimeter, ice cube trays, and turkey basters. A second box contained paper resources including guides for identifying microorganisms, field guides for plants and birds, and forms for recording data. These materials actually constrain student thinking in particular ways, guiding them toward science, although they are not aware of it. Because they know they could choose any other tools, students do not think that having a box with suggested resources limits what they can do. But as a consequence, they begin to see patterns in nature in terms of these instruments and paper resources.

The students developed a variety of experiments or plan observations. For their first experiment, John and Tim decided to find out whether there is a relationship between the speed of the creek at various places and its profile. They floated an orange in the creek and used a stopwatch to time how long it would take to float a five-meter distance, which they had determined using a measuring tape. They also used the measuring tape for determining the width of the creek at various places and determine its depth using a ruler. Jodie and his team wanted to determine if the number and type of organisms are related to the amount of oxygen available. With a dissolved-oxygen meter, they measured oxygen levels at four places in the creek. They used a one-square-foot Serber sampler to collect organisms. For each place, they classified and counted the organisms. Finally, they constructed a graph for each organism in which the counts were plotted against oxygen levels (abscissa). Michelle and her group began their work with a documentary. Equipped with a tape recorder and camera that they brought from home, they visited different sites along the creek. They took photographs depicting its state, including the various forms of garbage that they found. They recorded their observations on tape and transcribed them later for constructing a report, which they also presented at an open-house event in their community.

At various points throughout the science unit and at the end, each student group had opportunity to report the results of their diverse studies back to the entire class. In this way, everyone was aware of the studies that others were doing and learned about the science concepts involved. The teacher also invited students to comment on and criticize the work of others. Thinking about comments and critique allowed students to develop an understanding of what others had done. The children who participated in this curriculum learned more than just some science concepts disconnected from anything else they know. They learned about real phenomena and how science concepts from different disciplines were necessary for understanding the complexity of nature. In particular, this and other teaching units we studied revealed that open-inquiry laboratory activities allow students to develop higher-order process skills such as identifying variables, interpreting data, hypothesizing, defining, and experimenting. As with the biologists studying salmon retinae, these skills and the concepts students learn develop with their familiarity with the phenomena.

The students had come to understand science. But it was more important to the students that their work had a real effect on their community. Many citizens attended the open house and came to learn about the health of the creek from the students. More so, the results of their studies were also re-ported on the website of a local environmentalist group, and a local newspaper featured what they had done (see also "Scientific Literacy"). One student noted about the impact that the course had not only on her own learning but also on the community as a whole:

> During this course I learned about fieldwork: I learned how to collect samples of the creek and take temperatures and speed. I also did some work with the community. It taught me about working with others and working in the community. I noticed that ever since our Hagan Creek article was published in the *Peninsula News Review* that the public has begun to notice the creek. (Sally, June 1, 1998)

CONTROL OVER RESEARCH QUESTIONS, INSTRUMENTATIONS

When I began teaching school science, I believed that all students should engage in the same activity, use the same tools and instruments, and get the same results. My image for school science was based on the idea that students have to achieve one and the same specific result. I thought that if students conducted lab activities, they would see the science concepts. I soon realized that life did not work this way. I learned that one already needed to know the concept to recognize it in the phenomenon, like the biologists needed to know what the graph displayed to be sure about what the microscope displayed. I also learned that requiring all students to measure series of variables and represent the creek using graphs or histograms excluded especially girls and aboriginal students. While these students participated in the data collection, the subsequent data analyses and activities that focused on graphs, formulas, and other mathematical forms of presenting data generally turned them off. That is, not being in control of the experiment affected students' interest. It did not allow them to frame the motive of the activity.

Then I began to encourage students to define goals and investigate on their own terms, choosing their data collection and the ways to present results that best fit their interests and needs. All of a sudden, audio-recorded descriptions, videotaped records of the phenomena, photographs, drawings, and other ways of presenting phenomena began to proliferate. This change provided forms of knowing and learning that led to an increasing participation of previously excluded students; and through this participation and the periods of sharing, these students also came to access science. There exists considerable research evidence that shows that students who frame, conduct, and report their own research are much more interested and motivated, and, as a result, come to understand science in a much deeper way. Having taught and researched physics, chemistry, and biology at middle and high school levels, I know that students generally learn much more science content *and* about the nature of science when they engage in *authentic school science*. These students also remained interested in science, although they often did not pursue a related degree. Many of my students have taken the most advanced science classes (e.g., advanced placement) and then pursued studies in business, political science, or fine arts.

GETTING THE MOST FROM LABORATORY ACTIVITIES

We learn to do by doing. If we follow instructions, we learn to follow instructions. If we take tests, we learn how to take tests. If we design and conduct experiments, we learn to design and conduct experiments. To figure out how to get the most from laboratory activities, we need to determine what we want to teach and learn. If all we are interested in is to study for tests, which generally do not measure deep understanding, laboratory activities are not a good idea. If we want students to learn how to follow procedures, we use laboratory activities where every step is designed beforehand. In both these approaches, students risk to learn very little about how science describes the world that surrounds them. They do not learn how their own actions change not only the environment but also how they perceive it. Because learning concepts through laboratory activities takes a lot of time—also seen in the example of the biologists—it would not be wise to try to teach science concepts in this way if there is a shortage of time.

Doing open-inquiry laboratory activities, however, students develop scientific inquiry and technical skills, they learn how these are connected to specific scientific concepts, and they become adept at dealing with situations that are not clearly defined. They also learn to make do in situations of uncertainty, and they learn how to figure out things for themselves. That is, besides science, they learn about the nature of science. Finally, they also become independent problem solvers, just the kind of people who universities and employers are looking for.

Of course, any evaluation has to go hand in hand with the premises of the activity. Allowing students to do open-inquiry activities and then penalizing them if they do not get some specific result will do them a disservice. Students would fall back on asking us teachers what they wanted them to do. If students do *authentic* open-inquiry work at school, parents can assist by engaging their children in conversations about their projects. Parents might pose questions that will help their children develop their ideas about how to find out more. The children do not need answers from parents or others. They need, above all, support in figuring out ways of getting answers and perhaps support to access resources. Parents might be able to provide such resources so that their children could use them in the way Michelle had used the tape recorder and camera that she brought from home.

Science department heads and school administrators may have to support teachers, especially when they are timid and afraid that their students may not learn all they need for some high-stakes test. This support might come, for example, by making time so that science teachers (sometimes) teach together, learning from each other how to better support student inquiry. Some schools may be open to allow parents and others from the community to participate in science teaching, thereby making available additional learning resources (see also "Scientific Literacy"). Schools might also contact the various science and education departments of local colleges and universities to draw on the extant expertise or to entice science and science education students to do internships or volunteer work to support laboratory activities.

REFERENCES

Herron, M. Dudley. 1971. The Nature of Scientific Inquiry. *School Review* 79:171–212.

Knorr-Cetina, Karin D. 1981. *The Manufacture of Knowledge: An Essay on the Constructivist and Contextual Nature of Science.* Oxford: Pergamon Press.

Roth, Wolff-Michael. 1995. *Authentic School Science: Knowing and Learning in Open-Inquiry Science Laboratories.* Dordrecht, the Netherlands: Kluwer Academic Publishing.

Tobin, Kenneth. 1990. Research on Science Laboratory Activities: In Pursuit of Better Questions and Answers to Improve Learning. *School Science and Mathematics* 90:403–18.

ADDITIONAL RESOURCES

Suggested Readings

Hofstein, Avi, and Vince Lunetta. 2004. The Laboratory in Science Education: Foundations for the 21st Century. *Science Education* 88:25–54.

Latour, Bruno, and Steve Woolgar. 1986. *Laboratory Life: The Social Construction of Scientific Facts*. Princeton, NJ: Princeton University Press.

Websites

The Board on Science Education of the National Academies has published a practical guide to teaching inquiry as recommended by the *National Science Education Standards*: http://books.nap.edu/catalog/9596.html

The Exploratorium in San Francisco has authentic inquiry as one of its key aims: www.exploratorium.edu

The materials needed to do a unit on water quality at the elementary, middle, and high school levels are provided by the Washington Virtual Classroom: www.wavcc .org/wvc/cadre/WaterQuality/elementary_supplies_and_material.htm

Problem-Based Learning is an approach to curriculum that centers on real-life and interesting problems rather than content; the Illinois Mathematics and Science Academy has a range of interesting resources and links: www2.imsa.edu

7

Learning Environments in Science Classrooms

Barry J. Fraser

Students spend a huge amount of time at school—approximately 7,000 hours by the end of elementary school, around 15,000 hours by the completion of secondary school, and nearly 20,000 hours in educational institutions by the completion of a university course. Therefore, students certainly have a great interest in what happens to them at school and university, and students' reactions to and perceptions of their educational experiences are important. However, despite the obvious importance of what goes on in classrooms, most teachers and researchers rely heavily and sometimes exclusively on the assessment of academic achievement and other learning outcomes. Although nobody would dispute the importance of achievement, it cannot give a complete picture of the educational process.

This chapter is devoted to one approach to conceptualizing, assessing, and investigating what happens to students during their education. In particular, the main focus is on students' perceptions of important aspects of the learning environments of classrooms. Clearly, having positive classroom environments is a valuable goal of education. But, it should not be assumed that the equally important issue of student outcomes is ignored in this chapter. Extensive past research provides consistent evidence that the classroom environment is so strongly associated with student outcomes that it should not be ignored by those wishing to improve the effectiveness of schools and universities.

In contrast to methods that rely on outside observers, the most common assessment approach defines classroom environment in terms of the shared perceptions of the students and sometimes the teachers in that environment.

This has the two advantages of describing the class through the eyes of the actual participants and obtaining data that the observer could miss or consider unimportant. Students are in a good position to make judgments about classrooms because they have experienced many different learning environments and have had enough time in a class to form accurate opinions. Also, even if teachers are inconsistent in their day-to-day behavior, they usually give a consistent picture of the long-standing features of classroom environment. Although direct observation of classroom settings certainly is valuable, it doesn't tell the whole story about the complicated and subjective judgments made by students and others who influence learning.

BACKGROUND

A milestone in the historical development of the field of learning environments occurred approximately forty years ago when Herbert Walberg and Rudolf Moos began seminal independent programs of research that formed starting points of the work encompassed by this chapter. Walberg developed the Learning Environment Inventory as part of the research and evaluation activities of Harvard Project Physics, whereas Moos developed social climate scales for various human environments, including the Classroom Environment Scale. Herbert Walberg's (1979) and Rudolph Moos's (1979) pioneering work built upon earlier ideas of Kurt Lewin (1936) and Henry Murray (1938) several decades before. Lewin's field theory recognized that both the environment and its interaction with personal characteristics of the individual are potent determinants of human behavior. Lewin's formula, $B = f(P, E)$, stressed the need for new research strategies in which behavior is considered a function of the person and the environment. Murray proposed a needs-press model that allows the analogous representation of the person and the environment in common terms (Fraser 1998a).

Learning environments research originated in the United States, but soon spread to other countries, especially Australia (Fraser and Walberg 1991) and the Netherlands (Wubbels and Levy 1993). However, in the last decade, Asian researchers have made comprehensive and distinctive contributions (Goh and Khine 2002).

Schools and teachers often are subjected to judgments about their performance based only on simple measures of achievement. Certainly, students should be expected to demonstrate skills that can be measured. But this is only one of the objectives of education. Exclusive attention to achievement can detract from the human qualities that make education a worthwhile experience for students. Therefore, learning environment researchers over the years have looked for ways in which educational evaluation can be used more appropriately.

The curricula of schools and universities consist not just of content and outcomes—much as one might get that impression from listening to teachers, students, and their parents. They also consist of places, typically classrooms, where the business of learning takes place. Often it is the quality of life lived in classrooms that determines many of the things that we hope for from education. Also, the quality of classroom life is an important influence on the very achievement measures to which so much interest is directed.

ASSESSING LEARNING ENVIRONMENTS

Although classroom environment is a subtle concept, remarkable progress has been made in conceptualizing, assessing, and researching it. A considerable amount of work has been undertaken in many countries on developing methods for investigating how students and teachers perceive the environments in which they work. In particular, over the years, researchers have developed numerous questionnaires to assess students' perceptions of their classroom learning environments (Fraser 1998b). For example, these questionnaires provide information about whether a class is dominated by the teacher or is student centered; whether students actively participate in class or sit and listen to the teacher; whether students cooperate and discuss with each other when they are learning, or whether they work alone; whether the teacher is supportive and approachable; whether the students have a say in the choice of teaching and assessment methods; and whether differences in students' interests and speeds of working are allowed for by the teacher.

Some examples of classroom learning environment questionnaires, together with the dimensions that they assess, are given below (Fraser 1998b):

- *What Is Happening in This Class?* (WIHIC)—Student Cohesiveness, Teacher Support, Involvement, Investigation, Task Orientation, Cooperation, and Equity
- *Constructivist Learning Environment Survey* (CLES)—Personal Relevance, Uncertainty, Critical Voice, Shared Control, and Student Negotiation
- *Science Laboratory Environment Inventory* (SLEI)—Student Cohesiveness, Open-endedness, Integration, Rule Clarity, and Material Environment
- *Individualized Classroom Environment Questionnaire* (ICEQ)—Personalization, Participation, Independence, Investigation, and Differentiation

These questionnaires have been used in different countries and at different grade levels. They have been translated into various languages, including Spanish, Arabic, Chinese, Korean, Indonesian, Thai, and the South African language of North Soto. They have been used by hundreds of researchers, thousands of teachers, and millions of students around the world. Furthermore, most teachers find that it is easy and convenient to obtain information about their learning environments from their students. And this information can be useful for providing valuable feedback and for guiding attempts to improve teaching and learning.

RESEARCH ON LEARNING ENVIRONMENTS

Over the past few decades, learning environment researchers have attempted to answer many interesting questions. Does a classroom's environment affect student learning and attitudes? Can teachers conveniently assess the climates of their own classrooms, and can they change these environments? Is there a difference between actual and preferred classroom environment, as perceived by students, and does this matter in terms of student outcomes? Do teachers and their students perceive the same classroom environments similarly? How does the classroom environment change when a new curriculum or teaching method is introduced? Do students of different abilities, genders, or ethnic backgrounds perceive the same classroom differently? These questions represent the thrust of the work on classroom environment over the past thirty years (Fraser and Walberg 1991).

Researchers have carried out many dozens of studies into the relationship between student outcomes and the quality of the classroom learning environment (Fraser 1998a). These studies have been carried out in numerous countries and at various grade levels with tens of thousands of students. The consistent evidence from these studies is that the nature of the classroom environment is related to student outcomes (both cognitive and affective). Therefore, teachers should not feel that it is a waste of time for them to devote time and energy to improving their classroom environments. The research shows that attention to the classroom environment is likely to pay off in terms of improving student outcomes.

Classroom environment instruments have been used as a valuable source of process criteria in the evaluation of educational innovations. For example, in Singapore, the WIHIC was employed in evaluating adult computer application courses. Generally, students perceived their computing classes as being relatively high in involvement, teacher support, task orientation, and equity, but the course was differentially effective for students of different sexes and ages. In an evaluation of an urban systemic reform initiative in the United States, use of the CLES painted a disappointing picture in terms of a lack of success in achieving the constructivist-oriented reform of science education (Goh and Khine 2002).

Feedback information based on student perceptions has been employed in a five-step procedure as a basis for reflection upon, discussion of, and systematic attempts to improve classroom environments at various levels of education (Fraser 1986). First, all students in the class respond to the preferred form of a classroom environment instrument, with the actual form being administered in the same time slot about a week later (assessment). Second, the teacher is provided with feedback information derived from student responses in the form of profiles representing the class means of students' actual and preferred environment scores (feedback). These profiles permit identification of the changes in classroom environment needed to reduce major differences between the nature of the actual environment and that preferred by students. Third, the teacher engages in private reflection and informal discussion about the profiles in order to provide a basis for a decision about whether an attempt would be made to change the environment in terms of some of the dimensions (reflection and discussion). Fourth, the teacher introduces an intervention of approximately two months' duration in an attempt to change the classroom environment (intervention). Fifth, the student actual form of the scales (i.e., the environment the students perceive they actually are experiencing) is readministered at the end of the intervention to see whether students are perceiving their classroom environments differently from before (reassessment). These studies usually reveal that there has been an improvement in classroom environment and that teachers value their involvement in this action research aimed at improving classroom environments.

Although this chapter gives emphasis to assessing classroom environment through students' perceptions, which has been the predominant method in past research, it is important to note that significant progress has been made in using qualitative methods in learning environment research and in combining quantitative and qualitative methods within the same study of classroom environments (Tobin and Fraser 1998). For example, in a multilevel study of the learning environment, qualitative methods involved visiting classes, using student diaries, and interviewing a teacher-researcher, students, school administrators, and parents. A video camera recorded activities, field notes were written during and soon after observation, and team meetings took place regularly. Tobin and Fraser (1988, 639) conclude, "We cannot envision why learning environment researchers would opt for either qualitative or quantitative data, and we advocate the use of both in an effort to obtain credible and authentic outcomes."

CONCLUSION

Based on this chapter, the following implications emerge for improving science education:

1. Because measures of learning outcomes alone cannot provide a complete picture of the educational process, assessments of learning environment should also be used to provide information about subtle but important aspects of classroom life.

2. The evaluation of innovations and new curricula should include classroom environment instruments to provide economical, valid, and reliable process measures of effectiveness.

3. Teachers should use assessments of their students' perceptions of actual and preferred classroom environment to monitor and guide attempts to improve classrooms. The broad range of instruments available enables science teachers to select a questionnaire or particular scales to fit their personal circumstances.

4. When assessing and investigating classroom environment, a combination of qualitative and quantitative methods should be used instead of either method alone.

This chapter has tried to encourage others to use classroom environment assessments for a variety of research and practical purposes. Given the ready availability of questionnaires, the importance of classroom environment, the influence of classroom environment on student outcomes, and the value of environment assessments in guiding educational improvement, it seems desirable that researchers and teachers begin to include classroom environment in evaluations of educational effectiveness. Although educators around the world pay great attention to student achievement and only a little attention to the environment of school and university classrooms, research on classroom environment should not be buried under a pile of achievement tests. Hopefully, this chapter will encourage and guide important research and practical applications involving classroom environment.

REFERENCES

Fraser, Barry. 1986. *Classroom Environment.* London: Croom Helm.
———. 1998a. Science Learning Environments: Assessment, Effects and Determinants. In *International Handbook of Science Education,* edited by Barry J. Fraser and Kenneth G. Tobin, 527–64. Dordrecht, the Netherlands: Kluwer Academic.
———. 1998b. Classroom Environment Instruments: Development, Validity and Applications. *Learning Environments Research* 1:7–33.
Fraser, Barry, and Herbert J. Walberg, eds. 1991. *Educational Environments: Evaluation, Antecedents and Consequences.* Oxford: Pergamon Press.
Goh, Swee Chiew, and Myint Khine, eds. 2002. *Studies in Educational Learning Environments: An International Perspective.* Singapore: World Scientific Publishers.
Lewin, Kurt. 1936. *Principles of Topological Psychology.* New York: McGraw-Hill.
Moos, Rudolph H. 1979. Evaluating Educational Environments: Procedures, Measures, Findings and Policy Implications. San Francisco, CA: Jossey-Bass.
Murray, Henry A. 1938. *Explorations in Personality.* New York: Oxford University Press.
Tobin, Kenneth, and Barry J. Fraser. 1998. Qualitative and Quantitative Landscapes of Classroom Learning Environments. In *International Handbook of Science Education,* edited by Barry J. Fraser and Kenneth G. Tobin, 623–40. Dordrecht, the Netherlands: Kluwer Academic.
Walberg, Herbert J., ed. 1979. *Educational Environments and Effects: Evaluation, Policy, and Productivity.* Berkeley, CA: McCutchan.
Wubbels, Theo, and Jack Levy, eds. 1993. *Do You Know What You Look Like? Interpersonal Relationships in Education.* London: Falmer.

PART 2

LANGUAGE AND SCIENTIFIC LITERACY

8

Scientific Literacy

Wolff-Michael Roth

The concept of *scientific literacy* plays a central role in science education reform efforts. Educators generally agree that scientific literacy should be an important outcome of schooling. But despite its nearly fifty-year history, the term *scientific literacy* has eschewed a precise or agreed-upon definition. Paul de Hart Hurd, one of the founding fathers of the science education discipline, stipulated that a valid interpretation of scientific literacy must be consistent with the prevailing image of science and the revolutionary changes taking place in our society (Hurd 1998). For many science educators, this has meant that the adjective *scientific* refers to the science that scientists do. This does not have to be the case. There are educators who suggest that school science could orient itself with science as it is used and practiced in the everyday life of the communities we live in. This makes sense if we think that historically, elite and advanced forms of activity—science or professional science—always begin in their everyday equivalents.

ORALITY, LITERACY, SCIENCE

The *Oxford English Dictionary* explains the noun *literacy* and the adjective *literate* in terms of acquaintance with the written word and literature; this acquaintance is associated with being educated, instructed, or learned. Literacy contrasts with orality, the use of spoken language only, such as in the case in prewriting cultures. Although many people today mostly talk rather than read and write in their everyday lives, the fact that they are familiar with reading and writing shapes the way in which they think and work. More so, even though some individuals in Western societies cannot read or write—they are said to be *il*literate—they still think and go about their daily business in ways characteristic of a literate rather than an oral culture. Literacy and orality are different because the former permits a culture to reproduce itself using books and other media, whereas the latter has to rely on face-to-face transmission of knowledge, values, and beliefs. Furthermore, literacy allows new and more powerful forms of understanding to evolve, including science, mathematics, and technology. The things and tools that surround us in our daily life—including kitchen appliances, computers, electronic games, and cars—embody these forms of understanding. That is, the lives of those who participate in and are members of literate society are inherently shaped by science and technology. But this participation does not require knowing the concepts and theories of science or technology. We are able to drive cars and use kitchen appliances without having to know mechanical engineering or computer science, the domains of knowledge and expertise that contribute in the production of the modern car; and we need to know little about electromagnetism, knowledge of which is embodied in radios or kitchen appliances. We are scientifically and technologically literate to the extent that we use and learn to use artifacts that embody our culture's scientific and technological knowledge.

TRADITIONAL APPROACHES TO SCIENTIFIC LITERACY

For the purposes of conducting a large international study of science education, the Organisation for Economic Co-operation and Development defines *scientific literacy* in terms of peoples' capacities for using scientific knowledge, identifying questions, and drawing evidence-based conclusions. These capacities traditionally are thought to underlie any understanding and decision making about the natural world and the changes made to it by human actions. It is apparent that such a definition quickly leads us to think that science for *all* means every citizen has to be or think like a little scientist. Because most people do not know science in the way scientists do, some researchers take a deficit approach. They suggest that scientific literacy is a myth (e.g., Shamos 1995), and that as a whole we are too dumb to be scientifically literate. To support such arguments, these researchers point to a study that showed Harvard graduates who did not know that on the northern hemisphere, the earth was farther from the sun in the summer rather than in winter—the opposite of what one might expect about the relationship between the temperature and distance from a heat source like a stove. Or they point to the fact that many people have difficulties programming their VCR as evidence for lack of scientific knowledge. The authors of *Science Matters: Achieving Scientific Literacy* (Hazen and Trefil 1991), a book with widespread distribution, lists over 150 scientific words and names of scientists that everybody ought to know. There are entries about special relativity, Maxwell's equations, gauge particles, and other exotic and esoteric things that I myself have learned about in graduate studies of physics but that I have never had occasion to use in my everyday life. So what do we need to know or be able to do to be scientifically literate? Or, to foreshadow the examples below, what do we need to be able to do to make society as a whole more scientifically literate?

CRITIQUE OF SCIENTIFIC LITERACY FOR *ALL*

With hundreds of thousands of articles annually published in scientific journals, there not only exists an enormous body of scientific knowledge but also what scientists know grows exponentially. Yet any scientist only knows a fraction of all of this knowledge. What bits from among the existing scientific knowledge should students learn? It is evident that nobody can learn all of it. So which parts of science ought we to teach? Some may suggest that there are some basics that should be taught. But there is no evidence that we need to know, for example, Sir Isaac Newton's three laws of motion or Ohm's law of electricity to be successful in life, and even to be successful in most sciences.

Another critical issue in the question about scientific literacy for all pertains to the nature of science itself. Historically, mostly white men did science; they did not have to worry about other things in their lives, like making meals, bringing up children, or cleaning and maintaining the laboratories where they worked. They developed particular styles of doing and thinking, and legitimizing what they were doing and thinking. Because knowledge never just exists in itself but arises from experience, it is also tied up with our values. As a result, the forms of knowledge the scientists developed, too, are characterized by the way in which these scientists worked. Typically, they disconnected knowledge from everyday experience. Scientists, as we know so well from the history of the development of the atomic bomb, portray scientific knowledge as value free, objective (i.e., independent of the human subject), purely rational, universal, impersonal, unbiased, nonjudgmental, nondogmatic, and so on. Critics suggest that this approach is typical of white, middle-class, and male culture and does not represent the way that others—including women, aboriginal (or First Nations) peoples, African Americans, and working-class people—know (Harding 1998). In all these cultures, knowledge is experienced as relational; the object known is not separate from the subject who knows. This different way of knowing is quintessentially captured in *Feeling for the Organism*, the biography that Evelyn Fox Keller wrote about the Nobel Prize–winning woman scientist Barbara McClintock (Keller 1983).

As a result of their different everyday ways of knowing and scientific ways of knowing, women, First Nations people, and African Americans frequently experience a clash of cultures when they are faced with science that interferes with their learning and doing well in school science. The science educator Glen Aikenhead, who has worked a lot on understanding the differences in worldviews of science and other ways of knowing, suggested that the vast majority of students tend to become alienated by the differences between what they know and the culture of science. Trying to make everybody think like scientists is therefore a modern form of colonialism. Aikenhead suggests that we ought to teach science in ways that acknowledge the existence of other forms of knowing, for example, those that characterize the experiences of women, First Nations people, or African Americans. How, then, should we think or rather rethink scientific literacy?

RETHINKING SCIENTIFIC LITERACY

Some science educators propose to rethink scientific literacy in terms of the competencies of taking action (Roth and Barton 2004). However, taking action cannot be learned in the abstract, by reading books or memorizing some facts—reading about science no more makes us scientifically literate as watching sports on television makes us physically fit. Taking action is learned by taking action. It is when we read books and memorize facts to reach a meaningful goal that we come to understand science concepts and develop scientific literacy. More so, it is evident that we cannot simply engage in action but also have to take responsibility for our actions. Taking action therefore also implies ethics, morals, and values.

CONTESTED SCIENCE

One place where people should be able to take action is when it comes to their community and their interests. In the past, those in power often drew on scientists to deprive some members of society of their rights, or to push development ahead against the will of the people who inhabit the areas to be developed. Being able to stand up for and fight for one's rights and interests in contested issues that also involve science and scientists ought to be an important goal of scientific literacy for all. This does not mean that the people who stand up actually need to know about the structure of an atom; rather, those who argue for a rethinking suggest that *scientific literacy* refers to the competency to engage in the debate and, if necessary, to enroll specialists to assist them in their cause.

As a way of exemplifying this discussion, consider the following debate. It involves an engineer who had been hired by a town council in support of their attempts to avert the construction of a pipeline to provide a section of the town with the same water that everyone else already had. The engineer conducted some tests and wrote a report that was used to make the case against the pipeline extension. In part, the report suggested that the residents' well water was biologically safe, though it had some aesthetic problems (too many corrosive elements). However, the residents of the area began to protest. They were tired of having to replace their water pipes and appliances frequently; and every summer, they have to get their drinking water by driving three miles to the next gas station. Eventually, a public meeting was called. As part of the meeting, various scientists and engineers provided summaries of their report. The residents then could ask questions concerning the reports. In the following excerpt, one local resident (Knott) interrogated Logan, the chief scientist of Logan Engineering.

Knott: Would, do you know, for example whether chromium can be treated?

Logan: Yes, yes I do.

Knott: Successfully?

Logan: Yes, it can with ion exchange filtrate, a filter. I phoned the manufacturers of certain systems and they assured me that that can be done.

Knott: And that's good enough for you?

Logan: Well, I read it in publications as well.

Knott: Oh, there's a publication that we have here that says it has, that says there is no commercial treatment for chromium.

In his report, Logan had suggested that the residents should get filters to eliminate the chromium in the water. Unimpressed by the various university degrees that Logan was said to have, Knott stood up for his cause. He questioned the engineer's recommendations. Although Logan claimed that he had obtained information suggesting that water treatment for chromium existed, Knott pointed out that the information provided in the exhibition outside the meeting suggested that commercial treatment for the chemical currently was not available. Like Knott, other residents and concerned citizens queried the engineer. These queries showed that there were several possible errors in the scientific method Logan had used. More so, it became evident that Logan's assessment, taken at a moment when the water in the aquifer was about midlevel, wrongly represented those four months of the year when the well water could not be consumed.

The meeting also brought forth the tremendous knowledge and experience they had collected while they had lived in the area, which for some had been thirty years. There were residents who had hired their own consultants, and therefore had available scientific data for long periods of time. Residents also knew about the corrosion, which killed plants and destroyed plumbing and appliances. That is, although most residents had not studied science, they contributed to making scientific literacy happen in this public forum of the meeting.

How, then, might we help young people to become scientifically literate in the sense of being able to make scientific literacy happen in their community?

SCIENTIFIC LITERACY IN
DIVERSE LOCATIONS IN THE COMMUNITY

When we look at what people do in a community, we find them involved in many activities and in diverse locations. Some of these activities include scientists, engineers, and scientifically trained technicians. Other activities do not include them but still involve scientific knowledge. Science turns out to be but a fiber in the thread of social life. When we take a holistic look at any community, focusing on different groups and individuals, we generally find individuals and groups involved in producing knowledge and making changes to their environment. These changes often follow intense debates about possible ecological, economic, political, and social impacts. The contributions people make, the forms of knowledge they contribute, also differ—the residents in the previous example used the knowledge and experience that they had built up over a thirty-year period as a resource in their repeated attempts to get the water pipeline extended into their area. But in this same community, there were many other ways in which individuals and groups engaged; and they were successful, although nobody tested whether they knew the structure and composition of the atomic nucleus, or some other fact that scientists have accumulated and noted in articles and books.

In one instance, an environmentalist group raised the interest of residents to become stewards of the watershed and its creek. One important aspect of revitalizing the watershed was to make the creek suitable as the trout habitat that it had been when the white people arrived in this area. Suitable trout habitat has all the characteristics of a healthy stream—meandering channels, plenty of large woody debris and boulders, overhanging vegetation, cool temperatures, and high oxygen levels. Restoring a creek to trout-bearing capacity is also a move to restore many of its other aspects to what they would be if the stream was healthy. The environmentalists, some of whom had scientific or technological training, and stream stewards invited various experts to assist them in the project to walk the creek with them and to advise how to proceed with the restoration project. Various other groups also participated in and contributed to the creek improvement, such as university students who, as part of a summer job, conducted stream measurements and habitat assessments in and around the creek. It also included a student, who, as part of her master's project in environmental science, used a geographic information system to create multiple maps of the watershed and contributed the knowledge created back into the community. In the interactions with these others, everybody contributed to making scientific literacy happen in the community independently of whether they had studied science beyond the sophomore level in high school.

All of these activities constitute ideal contexts in which school students can participate and take action. The specific ways in which scientific literacy will be expressed across these settings will differ, because different social relations and material conditions will affect actions in different ways. Furthermore, these different social relations and material conditions will activate different dispositions in the participants, including students, again leading to different forms of knowledge that are brought to the situation at hand. For example, the students of a local middle school contributed to creating knowledge that was made available to environmentalists and the community as a whole (see also Chapter 6). Assisted in a variety of ways by their teachers, parents, elders, scientists, environmentalists, and others, some measured the relationship between the number of organisms and stream speed. Four girls documented pollution along different stretches of the creek using photographs and tape-recorded observations. One student had enlisted university scientists in his study of coliform bacteria counts in different locations, being able to relate high counts to specific chicken and cattle farms. Because the students designed research based on their own knowledge and interests, students who often come to hate science—girls and aboriginal students—were very involved in this unit.

Habitat assessment requires many decisions, which the students learned to arrive at by working with others and by using standard tools. They used a variety of forms as tools that allowed them to enter their estimates for the different dimensions that contributed to an assessment. These tools and forms provide constraints and structure student activity such that they are led to science rather than some other discipline. In the same way, assessing water quality could have been an insurmountable task had it not been for the variety of tools available for this activity. Students used a variety of scientific instruments. For example, the quality of the water in each reach was determined by testing temperature, dissolved oxygen, turbidity, and pH. They entered their readings in the water quality assessment form. This form combines different readings and calculations, and compares the outcome to an established calculation outcome to quality conversion.

Each year, the environmentalists organized an open-house event where the knowledge created and stream changes were reported. The stream stewards, university and middle school students, and many others contributed to the exhibitions in these well-attended events. The visitors included young people and adults alike, interacting with the various exhibitors. For example, the middle school students could be seen teaching children younger than themselves about how to check the health of the creek by taking stock of the microorganisms that live in various places. They also engaged in conversations with adults about how to identify these organisms or how to use the microscope for taking close-up looks at them. Somewhere else in the open house, two students explained why there might be different concentrations of microorganisms at different locations along the creek and what can be inferred from the different concentrations. At another location in the exhibit, a middle school student explains to an older, environmentally concerned citizen and activist how to operate a colorimeter, a device for measuring the turbidity of water. Around the wall of the room, there is an opened roll of graphing paper showing the water levels in the creek for an entire year; a water technician, hired by a farm situated on the banks of the creek, explains the different graph features. Elsewhere in the room, there is a three-dimensional model of the watershed that the environmentalist group had built for educational purposes.

As a result of these forms of participation, the conversations about the ecology of the watershed and Hagan Creek were extended to others, thereby extending scientific literacy to others in the community. Most importantly perhaps, the middle school students contributed to science and scientific literacy in the community. They did not just memorize bits and pieces of information, Newton's laws of motion, or the biological processes of meiosis and mitosis. These students actually contributed to making the community a different place. Through their part in the ongoing conversations about the ecological health of the creek and the watershed, these students did their part in maintaining and developing scientific literacy in their community. Viewed in this way, scientific literacy is something that exists at the level of the community, and what really counts is contributing to maintaining it, in whatsoever form the contribution might be. What individuals need to know to keep the conversation going is not quite clear at this moment in time. But it is evident that to maintain conversations, we need to be able to participate in them. It is through participation that we sustain conversations and, therefore, scientific literacy for *all*. It is a form of scientific literacy that is useful to society and therefore to each individual. Ultimately, in the case of the environment, such scientific literacy is useful to humanity as a whole.

SCIENTIFIC LITERACY AS COLLECTIVE PRACTICE

In the past, educators and educational policy makers thought about scientific literacy in terms of facts and stuff that people carry around in their heads. Educators tried to get as much of this stuff as possible into students' heads, which they could subsequently show off on exams. But they never or seldom needed to know any of this in their out-of-school and after-school lives. The example of scientific literacy and scientific expertise in the community gives us a different picture. *Scientific literacy* refers to the ability to competently and variedly participate in conversations where science is both a contested terrain—when residents fought the community politicians over the water pipeline—and resource—when the creek stewards sought the advice of biologists and water technicians for what to do to return their creek to its trout-bearing status. Although some of the stewards did not have a scientific training and most did not remember anything about their school science, all individuals participated in the conversations about enhancing the streams. These individuals participated in the environmental conversations as fibers of different material and color may contribute to making a thread. The fibers and the thread determine each other: the very existence of the thread depends on the presence of the fibers, but the value of the contribution each fiber makes to the whole depends on the thread. Science, once it is in the community, changes its nature because it has to make adjustments in its attempts to include and be accepted by the citizens. A scientific literacy that is truly democratic, truly is a science for *all*, cannot maintain its esoteric nature, in the same way that any other human pursuit, if it wants to open itself up, changes its nature as different people change it in the process of practice.

ABOUT BECOMING SCIENTIFICALLY LITERATE
AND DOING WELL ON TESTS

"How," some readers may ask themselves, "will participation in community affairs prepare me (or my child, or my student) for the tests that determine my (or my child's, or my student's) future? Do we not have to prepare directly for the tests?" I taught science for many years allowing students to choose topics and means of activities based on their interests but helping and directing them in the process. I know that my students learned a lot of physics or biology. As a teacher, I provide a tool box and paper resources (such as science books and field guides), which inherently structure student activity as soon as they begin to see them as resources. Students will be inclined toward the scientific, because they use and think in terms of the tools, instruments, and paper resources available. As a teacher and parent, my highest priority always has been to educate students for life. The contents of tests and processes of testing may be inconsistent with this priority. But I found that many students came to know not only what the curriculum prescribed but also the contents of the next, more difficult level of the subject. That is, learning with others and to achieve specific purposes—such as contributing to a community's collective knowledge—is not only fun but also enormously efficient.

In the course of my career, I moved to allowing students learn in the way I had done in the environmental unit (see also Chapter 6). I would always teach in this manner, attempting to interest students in finding out more and from different perspectives. I realize that students would not learn the same concepts at the same time. Therefore, to get my students ready for any type of exam, I would spend a sufficient number of lessons to prepare them for the specific question formats that they can expect on the particular test to be taken. As a parent, I would do things with my children in the community and other science-related activities that are truly meaningful and engaging to them *and* would assist them in studying for the formal tests. I would also see if I could get involved in assisting science teachers in the way that the parents of the students in the environmental unit did. As a science department head, I encouraged other teachers to teach through engagement in useful and meaningful activity—as a way of introducing other science teachers, they taught with me and I taught with them. As a department head or administrator, I would continue to encourage teachers to teach in this way and to set aside the required time to study for tests and acquisition of test-taking skills.

REFERENCES

Harding, Sandra. 1998. *Is Science Multicultural? Postcolonialisms, Feminisms, and Epistemologies.* Bloomington: Indiana University Press.

Hazen, Robert M., and James Trefil. 1991. *Science Matters: Achieving Scientific Literacy.* New York: Doubleday.

Hurd, Paul de Hart. 1998. Scientific Literacy: New Minds for a Changing World. *Science Education* 82:407–16.

Keller, Evelyn Fox. 1983. *A Feeling for the Organism: The Life and Work of Barbara McClintock*. San Francisco: W. H. Freeman.

Roth, Wolff-Michael, and Angela Calabrese Barton. 2004. *Rethinking Scientific Literacy*. New York: Routledge.

Shamos, Morris H. 1995. *The Myth of Scientific Literacy*. New Brunswick, NJ: Rutgers University Press.

ADDITIONAL RESOURCES

Suggested Reading

Roth, Wolff-Michael, and Jacques Désautels, eds. 2002. *Science Education for/as Sociopolitical Action*. New York: Peter Lang.

Websites

The Association of American Colleges and Universities has a website dedicated to the question how to make science more attractive to women: www.aacu-edu.org/womenscilit/index.cfm

The North Central Regional Educational Laboratory has a page with links to resources in scientific literacy: www.ncrel.org/engauge/skills/scilit.htm

Texts about a multicultural approach to science education can be found on the web page of Glen Aikenhead: www.usask.ca/education/people/aikenhead/

9

Verbal and Nonverbal Interactions in Science Classrooms

Kenneth Tobin

As Angela Calabrese Barton describes in Chapter 24, affording the agency of participants through science education is a central role of science educators. Within schools it is important that science curricula are enacted in ways that allow students to have the power to act to pursue their goals while surpassing the requirements for school graduation. As Gloria Ladson-Billings (1994) noted in her description of exemplary mathematics classes, the learning of African American youth is facilitated by a sense of *all for one and one for all*. That is, while participating in a class, all participants watch out not only for their own learning but also for the learning of others. In an important sense, the phraseology of *all for one and one for all* captures the essence of the dialectical relationship[1] between agency and structure, that is, between the power that individuals and collectives have to act and the resources they can appropriate to support their actions.

The verbal interaction that occurs in a classroom is a central structure that often makes or breaks the quality of the learning environments. At their best, verbal structures can facilitate learning. Listeners can make sense of others' utterances, connecting what they hear to their own understandings and reconciling contradictions, as they arise or at some later time. Speakers represent what they know as talk, and in so doing they make it an object for review—by themselves and others. Speaking, as cultural enactment, fits within a network of interactions and is a structure for others to appropriate. Accordingly, by speaking, a participant learns to enact talk in ways that are relevant, timely, and fluent. Verbal interaction is a cultural act, and the extent to which it is successful depends on whether or not the enactment of turns at talk leads to a community meeting its goals. It is never a question of an individual meeting her goals or collective goals being accomplished, rather; for talk to be regarded as successful, both individual and collective goals should be afforded. Hence, the quality of verbal interaction must be evaluated against individual and collective markers of success.

VERBAL INTERACTION AS A STRUCTURE

If a teacher is going to facilitate the learning of her students, she must interact with them successfully, acting in ways that provide all students with resources to support their learning. Through her actions, verbal and non-verbal, the teacher's practices are resources for all participants to use as a basis for meeting their goals. That is, a teacher's practices are part of a dynamic and unfolding structure that is dialectically interconnected with the agencies of all participants. The teacher's practices can be appropriated consciously and unconsciously as interactions occur continuously, creating a dynamic structural environment that can expand collective agency and associated practices. Constant changes in structure imply that agency also is changing continuously, creating resources for all participants to get involved and thereby contribute to the learning of self and other.

Consider the example of a teacher question. If a teacher asks students to describe the characteristics of a strong acid, the question can be tackled in many different ways. Different students might think about alternative answers, and the teacher may consider possible follow-up verbal moves contingent on whether or not correct, partially correct, or incorrect students' responses are provided. Having formulated possible responses, a student can decide whether or not to volunteer a response or to ask a question requiring the teacher to clarify, elaborate, or perhaps ask a different question. With every action or failure to act, the possibilities unfold in ways that are anticipated by most participants—anticipation leading to increased prospects for successful interactions. It is not just the words of the question that are resources; so too are the pauses, intensity, and pitch of utterances. Together these features of speech contribute to rhythm and verve, characteristics that get listeners in synchrony (or out of synchrony) with a speaker and elicit active forms of participation. For example, a fluent and lively speaker might elicit trains of thought in the listening audience, head nods, and occasional verbal affirmations of agreement or disagreement (e.g., *uh huh, mmm, yep,* and *no way*). A long pause following a question might elicit thought and efforts to frame responses, or, depending on the circumstances, a listener's thoughts might wander to consider other matters. Also, long pauses might catalyze a change of speaker. Accordingly, whether or not a pause is used for deep thinking on the subject matter to be learned depends on how active the listener is when the pause occurs. Contrary to the early research on wait time,[2] which seems to argue that longer pauses are universally beneficial, like any other resource, a pause can be appropriated by participants as they pursue the goals that have priority in the moment. Of course, if a student doesn't know what to do or how to do it, a pause intended to elicit deep thinking can add to frustration and produce other negative emotions.

A sequence of verbal interaction that occurs frequently in whole-class teaching consists of a teacher question (i.e., initiation—I), a student response (R), and a teacher evaluation (E) of the adequacy of the response. IRE patterns are prevalent and often are regarded as undesirable. There are good reasons for IRE patterns to be a cause for concern; in most science classrooms, they involve providing answers to questions when the correct answers are known, at least by a teacher and usually by most of the class too. Most questions call for facts, the teacher gets either one- to two-word answers or is left to answer his own questions, and the evaluations tend to be very short and nonsubstantive. In these scenarios, IRE sequences appear to contribute to poor learning environments. Two options appear viable when IRE patterns are ineffective—improve the quality of IRE interaction chains and/or reduce their incidence.

What makes an effective IRE interaction chain? Presumably the question, or the verbal move that initiates an interaction chain, is intended to transfer a turn at talk to students. To be regarded as successful, a student response would have to be in synchrony with the teacher's initiation and not breach the flow of the lesson. Hence the pause between I and R would allow students to formulate appropriate responses without losing momentum. Just how long the pause should be depends very much on the nature of the initiation. For example, if a teacher is involved in having students learn chemical symbols for common elements, she does not want a lengthy pause after her question; in fact she might regard any discernible pause as too long. Hence an IRE taking the following form might be regarded as successful.

Teacher: Calcium↑

(0.1s)

Connie: Ca.

(0.1s)

Teacher: Good.

The initiation consists of one word, with a rising inflection denoting a question (↑). The pause is barely discernible and a student provides a correct answer, which is affirmed quickly by the teacher. The rapid-fire IRE sequence is then repeated for different elements twenty times, after which the teacher enacts a similar drill on atomic masses of elements. In a context of the goals of this part of a lesson, the prevalence of IRE sequences consisting of (1) a short wait time that is almost immeasurable; and (2) one-word utterances might be regarded as appropriate, especially if facts are to be memorized and retrieved on demand. Similarly, the IRE pattern observed in the following example is considered fitting.

Teacher: It seems as if there is a higher incidence of asthma among the inhabitants of high-rise apartments in cities like ours. What are some of the likely causes?

(3.4s)

Jamal: I have read that it is possibly due to people eating their meals on their beds as they watch TV. The roaches then come to eat the small scraps of food left on the bed. People can then breathe in roach dung as they sleep.

(2.7s)

Teacher: That sounds like a promising start. How would roach dung contribute to asthma? Anyone?

In this example, the average wait time is about three seconds, the question required and received a thoughtful answer, and the evaluation provided support for the student who answered and referred a follow-up question to the class—hence continuing the IRE chain. Provided that IRE sequences such as this do not extend for too long, they seem quite appropriate, focus talking and thinking, and transfer turns at talk.

In considering alternatives to a high incidence of IRE interaction chains, it is important to consider the purposes of verbal interaction. When planning ahead it is prudent to focus on the goals of a teaching episode and enact verbal interaction accordingly. If there is talking, then it is assumed that listeners will engage actively and remain engaged for as long as the talk lasts. If this happens, and interest does not wane, the talk will likely expand the agencies of participants. However, lengthy monologues from a teacher can be perceived as boring and lead to frustration, inattention, and perhaps efforts to disrupt the flow of the teacher talk. Hence, whether or not to distribute the turns at talk and the types of talk evenly across participants depends on the purposes and the ways in which students are organized.

GETTING THE BEST FROM WHOLE-CLASS INTERACTIONS

If a teacher's verbal utterances are to be effective, they should engender participation of others through active listening. That is, students should make sense of utterances, and follow along as the dialogue continues. Listeners should fluently make sense of what is being said, presumably without having to ponder the meaning of words and phrases as the dialogue progresses. If students take a time-out to make sense of a phrase, they miss the continuing talk and the sense-making process is breached. Accordingly, characteristics of speech such as clarity and comprehensibility are critical if listeners are to make sense of and keep pace with the unfolding dialogue. Several factors are critical in making sense of talk. Pacing needs to be in tune with the sense-making capabilities of learners, which will vary within a class, often dramatically. Imagine a student who encounters a difficulty in making sense of the term *strong acid*. As he ponders the meaning of *strong*, the verbal interactions in the classroom continue, and his frustration mounts when he encounters other terms and phrases that he does not understand, such as *concentration of hydrogen ions, pH,* and 10^{-2}—and as the dialogue progresses fluently, frustration can be accompanied by feelings of failure. Clearly this is a scenario most teachers, students, and parents want to avoid. A student is left behind because of his failure to make sense of chemistry as it unfolds.

As Beth Wassell points out in Chapter 13, students for whom English is not a first language are unlikely to fluently make sense of instruction in English without having their participation breached by lack of understanding and associated moments of uncertainty. In these circumstances, it would be useful for the teacher to know who is and who is not making sense of what she is saying. Some teachers ask students to give a thumb down sign when they do not understand. This allows tutors to notice when help is needed. As a general rule, whole-class interactions might be curtailed in length and frequency of occurrence so that students who cannot follow fluently are not too disadvantaged. Annoyance and frustration might be signs that a speaker has talked for too long. Hence, teachers can monitor the verbal and nonverbal landscape of the classroom for indicators of negative emotions and act to minimize them.

In classes with a high proportion of students with limited English proficiency, students in need of assistance might be seated adjacent to peer tutors who can provide them with assistance during whole-class interactions. I have used peer tutors from the class; high-achieving students who can easily follow the unfolding dialogue and are willing to tutor others continuously. If tutoring is to occur in whole-class settings, it should be negotiated with students as a legitimate form of activity that is enacted quietly, to avoid breaching participation in mainstream activity. In addition to peers from the same class, I have used an array of tutors, including coteachers, new teachers seeking certification, faculty from institutes of higher education, and students who have already passed the course and are gaining elective credit through peer tutoring. My experience so far suggests that both the tutor and the tutee benefit from tutoring, and distractions to the teacher and other students in the vicinity are minimal.

When teachers talk for lengthy periods of time, they can lose the students and create negative emotional energy in the classroom, evident in signs of widespread dissatisfaction. If the purpose of whole-class interactions is for all participants to follow and learn, then it is important that speakers are fluent, comprehensible (hence clear), and relevant. Breaches of the speaker should not occur, nor should there be breaches in the sense making of listeners. This implies that exchanges of speaking turns are smooth and that all speech is audible and appropriate (i.e., engaging).

Minimizing the duration of whole-class interactions has substantial merit. Mary Budd Rowe (1983) did research on the 10-2 method. What this involves is breaking down whole-class interactions so that in every ten-minute interval, there is two minutes of small-group discussion on what has transpired in the previous eight minutes. Small-group participation can be structured by

questions from the teacher, requiring students to discuss and formulate responses by interacting with their nearest neighbors. "Nearest neighbor"
small-group discussions are most successful when the highest achieving
students are distributed around the room, able to participate as leaders in
discussion, and clarifying and elaborating on what was taught earlier. Hence
the two minutes of discussion provide a chance to query, clarify, elaborate,
evaluate, and tutor. The teacher can also monitor what is happening classwide
or participate in tutoring of students as the need arises. If the small-group
activities are successful, they might flow for more than two minutes and
students also might have additional resources to appropriate as they learn
from one another (e.g., reference books, charts, Internet sources, and DVDs).
However, for teachers and students who are unaccustomed to small-group
discussions, the 10-2 strategy is an effective way to learn how to learn and
enact new roles in small groups.

The forms of science talk that are appropriate depend on the purposes of
whole-class interactions. If new material is introduced, it makes sense to have
explanations and coordination of speech with gestures and material resources
such as diagrams, charts, and photos. Intensity, pitch, and pauses serve to
emphasize words and phrases and engage listeners. Questions might be
anticipated and encouraged by listeners when explanations do not make
sense, and, based on the teachers' scanning of nonverbal signs, teacher
questions might be directed to particular listeners to check on their comprehension or attentiveness. In such cases, students would respond and some
evaluation might be expected on the adequacy of their responses. During
whole-class interactions, there may or may not be exchanges of turn taking—
depending on the purposes of the activity. However, levels of attentiveness
seem to fall off when individuals are required to limit their participation to
active listening. Remaining active over lengthy periods of time can be fraught
with problems.

VERBAL INTERACTIONS IN SMALLER GROUPS

In small groups and one-on-one verbal interaction, the focus can be on
success. If interactions focus on the attainment of participants' goals, as
progress is made, those involved should experience positive emotions associated with their successes. My research has explored the synchrony of actions
making up an interaction chain. As a teacher interacts with students, I examine the body orientations of students and the teacher, movements of their
bodies (especially the head and eyes), facial expressions (such as smiles and
frowns), and gesturing. Similarly I examine whether there is alignment in the
speech patterns, including rhythm, intensity, pitch, and the distribution of
pauses. When participants are in synchrony, their interactions often produce
successful outcomes and positive emotions. Conversely, signs of asynchrony
are often associated with lack of success and negative emotions, such as

annoyance, frustration, disappointment, and loss of self-esteem—signs of a need for a change in roles of the participants. Teachers and students can use emotional climate as a barometer for the health of learning environments and fine-tuning their roles.

When there are high levels of synchrony in a group, the patterns of speech might have features not usually associated with what has traditionally been regarded as high-quality verbal interaction. For example, a person's head nods might signal that she understands what a person is saying and where she is going. In such an instance, there might no longer be a need to complete what is being said and the speaker might trail off, leaving an incomplete sentence. Alternatively, the speaker might understand the head nods and pause, thereby providing a resource for the listener to take the speech turn and complete the sentence. Hence, rather than an interruption (which is often associated with negative emotions), taking a turn at talk in this way may be a sign of coparticipation among two or more class members—indicative of successful interactions.

It is important that participants monitor the interactions within a group and from time to time identify contradictions and create plans to eliminate them. One common sign of dysfunction is the *turn hog*, a participant who speaks for excessively long periods of time or who takes a disproportionate number of turns at talk (can a teacher be a turn hog?). Within a small group, it is assumed that taking a turn at talk is an important form of participation, an opportunity to describe, evaluate, and question. Accordingly, among the factors to be reviewed by a group are the distribution of turns, the average duration of talking turns, and the distribution among participants of questions, explanations, and evaluations. Given my suggestion that participants monitor the emotional climate, it also makes sense to systematically review indicators of emotional energy by explicitly seeking feedback on positive and negative emotional markers, such as enjoyment, interest, challenge, frustration, anger, and resentment.

It is not wise to assume that students know what to do to initiate and sustain productive group discussions. Accordingly, it may be prudent for teachers to begin with very small groups that convene for a short time to complete an assigned task. For example, groups of two or three could discuss answers to given questions and seek to reach consensus on what is agreed and what contradictions need to be resolved. An initial (simple) rule structure would be to share turns at talk (speaking time), and alternate the roles of group leader, recorder, and reporter. As students learn the roles of speaking to benefit learning and attentive listening, the time spent in small groups can be increased, as can the number of participants per group.

In the 1970s and 1980s, there was a pervasive idea that teachers would participate in small groups (especially) as formative evaluators of what students were doing. For example, in her landmark book, *Teaching Science as Continuous Inquiry*, Mary Budd Rowe (1973) advocated that teachers join small groups without comment, sit so that their head is at the eye level of students, and listen attentively to what is happening in the group. If the group is productively engaged, the teacher might move on to the next group without participating verbally. Accordingly, there was a great deal of monitoring as teachers moved from group to group, evaluating progress through attentive observation. In contrast, some of the more effective teachers in urban science classes, described in our recent 2005 publication entitled *Improving Urban Science Education: New Roles for Teachers, Students and Researchers* (Tobin, Elmesky, and Seiler 2005), assumed very central and active roles whenever they joined a small group. For example, when Cristobal Carambo taught small groups, he was a central participant, doing science when he joined each group, and allowing students in the group to experience how a scientist (qua teacher) does science and to learn at his elbow, usually by participating in peripheral ways. Accordingly, his time with a group tended to be "for as long as it took," and his involvement was substantive. For central roles like these to be effective, it is essential that all students know what to expect of their teacher, accept the teacher's active role when he joins a group, and be aware of their own possibilities for participation.

BECOMING FLUENT IN TALKING SCIENCE

The use of language to interact in fields in which science is enacted is an essential part of doing science. Often it is necessary to enact talk fluently so that the talk is relevant, timely, and anticipated. If these criteria are met, then all participants can anticipate what is said and use it in the pursuit of collective goals. However, in order to become fluent in talking about science, it is necessary to rehearse, with others and alone. Next, I provide examples of rehearsal strategies to build oral fluency in science.

Imagine you are called as a student to testify to a federal judge on the necessity for all science teachers to make the following statement at the beginning of a ninth-grade course on living environment: "Darwin's theory of evolution is just a theory. There is an alternative theory called intelligent design. There are books on intelligent design in the library that you are encouraged to read if you want to learn about this theory." What would you say to the judge? Plan a speech, practice it, and anticipate the questions the judge might ask. Also, practice answers to the questions.

Although students will likely make notes as they plan for this activity, they should also rehearse alone, with parents or siblings, and go over the parts of the testimony in their heads. Rehearsing aloud with others and alone is an important step in the process of using language to represent science fluently. Homework exercises such as these can be developed by teachers, parents, and the students, and can be referenced to the science topics being taught at the present time. For example, a rehearsal scenario related to an upcoming lesson on motion might take the following form:

Rehearse your answers to the following questions your teacher is likely to ask in today's lesson on motion.

Why is it dangerous to place heavy objects on the back shelf of a car?
Why should bicycle riders wear safety helmets?
Is it safer for boxers to fight with or without gloves?

In each case, anticipate follow-up questions your teacher is likely to ask and rehearse your answers to these questions.

LEARNING TO INTERACT ACROSS CULTURAL BORDERS

To the extent that verbal interaction is a cultural activity, the challenge for science education can be cast in terms of adapting and aligning the talk from different cultures. Although students may attend the same school and be grouped together in the same science class, their histories may differ profoundly from one another in terms of race, class, and language resources. In many, and perhaps most, science classes in the United States, these categories are one way to express the diversities associated with teaching and learning science. The issue of diversity among students is only compounded by the diversity among the science teaching workforce and the way this becomes manifest in terms of cultural differences between the science teacher and most science students. Effectively communicating in a context of cultural diversity is the way I have thought about and written this chapter. There are many issues and ways to look at them. Central in all of them is the way participants talk to one another and make sense of what is said. From the perspective of a teacher, there are gaps in age; and successfully interacting with children, breaking down differences in child, youth, and adult culture, necessitates practice and sensitive listening. Of course, it is not just teachers who have to learn to adapt to and align culture, but the students as well. Hence, if science education is to flourish, the young students and the older teacher must learn to effectively interact.

Setting up discussions between pairs of students in science classes is an opportunity for them to learn to interact successfully across such boundaries. Such a grouping arrangement is also a wonderful opportunity for all students to talk science regularly and to become fluent. By rotating the membership of discussion pairs, students can learn to adapt their culture with peers who are culturally other, due to differences in race, gender and possibly class.

I have found one-on-one tutoring to be an ideal context in which teachers and students can learn how to build synchrony with one another when they differ in age, race, class, and sometimes gender. Also, as is described by Sarah-Kate LaVan (Chapter 31), Stephen Ritchie (Chapter 32), and me (Chapter 1), cogenerative dialogues can be effective in resolving problems and producing cultural alignment between teachers and students as small groups of students (usually selected for their differences from one another) interact with one another and the teacher to identify contradictions from their science classes and ways to resolve them. Working together to resolve problems and enact changes in science classrooms is a superb way to learn how to interact successfully with youth from different backgrounds of race, class, and gender. Not only do participants learn new culture that can be used in science classrooms to improve the achievement and participation of all, but they also build solidarity—a sense of belonging and of assuming responsibility for the success of all participants in their class. What is learned from participation in cogenerative dialogues and changes that are agreed upon can be enacted in science classes, leading to improved opportunities for learning science.

In many instances social class is an issue that mitigates the success of verbal interaction, especially in inner-city schools—and often differences in class are confounded with race. Growing up within a particular class and race provides opportunities to build distinctive forms of culture on which all learning can grow. However, if a teacher has not experienced these forms of culture, growing up in a different class and having a different race, then interactions may break down because of a failure to interact successfully with many students. In this chapter, I urge science educators to set up structures whereby students and teachers learn to interact successfully in groups ranging from one-on-one to whole class, so that all participants understand one another culturally and are able to adapt and align, producing successful interactions, positive emotional energy, and learning for all. I close with a word of caution. The necessary alignments and adjustments will not happen automatically in most science classes, and, if nothing conscious is done to create and maintain successful interactions, the status quo will likely be reproduced—a status quo replete with oppression and inequity. The hopes embodied in science education, for expanded agency and improved social trajectories, are contingent on cultural others successfully interacting to produce forms of science education that embrace a spirit of *one for all and all for one*!

NOTES

1. One way to think of a dialectical relationship is *both/and* rather than *either/or*. That is, the two parts of a seeming dichotomy constitute a whole.

2. Pauses between speakers and within utterances.

REFERENCES

Ladson-Billings, Gloria. 1994. *The Dreamkeepers: Successful Teachers of African American Children*. San Francisco: Jossey-Bass.

Rowe, Mary B. 1973. *Teaching Science as Continuous Inquiry*. New York: McGraw-Hill.

———. 1983. Getting Chemistry off the Killer Course List. *Journal of Chemical Education* 60:954–56.

Tobin, Kenneth, Rowhea Elmesky, and Gale Seiler, eds. 2005. *Improving Urban Science Education: New Roles for Teachers, Students and Researchers.* New York: Rowman & Littlefield.

ADDITIONAL RESOURCES

Cazden, Courtney B. 1988. *Classroom Discourse: The Language of Teaching and Learning.* Portsmouth, NH: Heinemann.

Lemke, Jay L. 1990. *Talking Science: Language, Learning and Values.* Norwood, NJ: Ablex.

Roth, Wolff-Michael, and Kenneth Tobin, eds. 2002. *At the Elbows of Another: Learning to Teach through Coteaching.* New York: Peter Lang.

———, eds. 2005. *Teaching Together, Learning Together.* New York: Peter Lang.

Tobin, Kenneth G. 1987. The Role of Wait Time in Higher Cognitive Level Learning. *Review of Educational Research* 57:69–95.

10

Understanding Visuals in Science Textbooks

Lilian Pozzer-Ardenghi and Wolff-Michael Roth

Images are pervasive in our culture. Television, video games, computer animations, and advertisements around town are examples of visuals that are integral parts of our daily lives. Moreover, the lives we lead today are made possible by science and technology, which came into being because of and simultaneously with the visuals. It is therefore not surprising that school science textbooks are full of visuals to illustrate concepts treated in the texts. However, visuals should and can do more than merely illustrate something. Research shows that they can be resources that assist students in understanding scientific concepts, especially when these concepts are unfamiliar or abstract. To achieve this potential, however, visuals need to be properly employed by textbook authors; most importantly, teachers and students need to develop knowledgeable ways of reading and interpreting them. While every student, parent, and teacher knows that being able to use language to communicate in and about science is important, there are few people who have realized the great degree to which visual literacy can contribute to doing and understanding science. Here we focus on the visuals that appear in science textbooks to help readers in better understanding the potential of this mode of communication, and we provide examples of how to work with photographs and line graphs to make better use of them for sense making and learning in science. Students, parents, and teachers may find our exemplary ways of reading useful for themselves or for helping others read visuals more critically and in greater depth, thereby realizing the potential of images to learning and doing science.

FROM VISUALS TO REPRESENTATIONS AND INSCRIPTIONS

Visuals are used to stand for something else. Thus, a photograph of a wedding stands for the wedding event; it makes the wedding present again, *represents* it even many years after it actually occurred. In the past, researchers have referred to visuals as *representations*. But we can also create visual images of the wedding in our minds when we remember it, and psychologists and cognitive scientists refer to such visual images as *representations*. There is therefore a potential confusion: does the word *representation* stand for something we envision in our minds or a visual that is external to ourselves? Sociologists and anthropologists have therefore begun to use an alternative word for representation: *inscription*. The term *inscription* already implies that the visual is materially inscribed in some medium—for example, on paper or on a computer monitor—and therefore cannot refer to mental content.

Inscriptions come in a great variety. Some of them resemble the things they stand for: a *photograph* of a rose resembles the rose we have seen, but someone looking at the *drawing* of a rose and comparing it to the roses in a garden will notice that it depicts more than one rose and therefore refers to roses more generally (see Figure 10.1). That is, there is a distance, a gap, between the drawing of the rose and the roses we see in the garden that is a bit larger than the distance between the photograph and the actual rose. Even less concrete detail exists in *diagrams* of roses used in biology to teach its parts; in this case, the diagram of the rose (Figure 10.1) refers to *all* roses. Scientists may be interested in the growth of the rose. They might take measurements at different times of the year and then produce a *graph* that shows the average height of roses plotted against time of the year; this inscription also refers to roses, but no longer has a likeness with them. Finally, the scientists might come up with an *equation* of rose height as a function of time, such as "$h = k \cdot t$."

As we move through the sequence of these different inscriptions representing a rose or something about roses, we notice that they are characterized by a

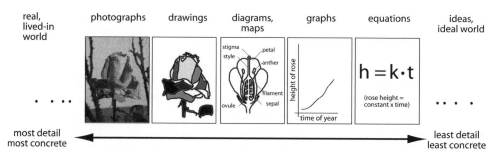

FIGURE 10.1 Inscriptions can be ordered along a continuum from *most concrete* (most detail and most likeness with the real, lived-in world) to *least concrete* (least detail, with no likeness with the real, lived-in world). The further we move to the right in this continuum, the more general and abstract the inscription, and the more information is "packed" into it.

decreasing amount of detail and concreteness but an increasing amount of abstractness and generality. That is, inscriptions can be characterized along a scale (see Figure 10.1) that has at one extreme the real, lived-in world in its fullness and particularity, and on the other extreme the pure idea, unsullied by any concreteness whatsoever. Importantly, the distance of an inscription from the real, lived-in world is proportional to the amount of work that the reader has to perform to understand what it is about.

PHOTOGRAPHS

A famous adage states that a photograph is worth a thousand words. Photographs are full of details that make them realistic but, at the same time, may cause confusion when interpreting them; in fact, a photograph may be worth a lot more than a thousand words and give rise to many different interpretations. When used in textbooks, photographs function as illustra-

FIGURE 10.2   The caption of this photograph is very brief and does not help readers to identify which of the many different plants are the fennel plants.

tions of a particular phenomenon or object to which the text refers. However, some textbooks present photographs without the necessary resources that would help students to associate the photographs with the text. For example, although most photographs come with captions, some photographs do not. In these situations, students have difficulty identifying what the photograph is an illustration of. Consider Figure 10.2. How do you start interpreting this photograph? For example, upon reading the caption, you may try to identify the fennel plants in this photograph.

You start by noticing that there is one type of plant with big leaves in the left-upper corner of this photograph, a different kind of plant in the center-right area, and one isolated little plant near the bottom-left corner. Because of conventions of perspective, your are able to interpret the plants on the left-upper corner of the photograph as being *behind* the plants in the center-right area, and the small plant in the bottom-left corner as being *in front of* all the other plants. But which of these plants are the fennel plants? You may notice that it is not possible to see the entire plant on the bottom-left corner of this picture, so this one probably is *not* a fennel plant. Similarly, the plants in the *back* (upper-left corner) of the picture are a little out of focus, so you now are left with the plants in the center of the photograph, the ones that can be clearly and fully seen. However, unless you already know from experience what a fennel plant looks like, you cannot be completely sure that these are the fennel plants. Other resources would be necessary to help you identify the fennel plants.

Consider now Figure 10.3. The photograph shows a salamander; there are a caption and a main text accompanying it. The title, "Salamanders," already tells you what the text will probably be about. As you read through it, you find more

Salamanders

Besides frogs and toads, **salamanders** (also called newts) belong in the order *Amphibia*. Salamanders can be found in the Americas and in the temperate zones of Asia, Europe, and Northern Africa. These animals look like lizards, but their skin is very soft and moist, and they do not have scales or claws. They are mostly carnivorous, eating insects, larvae, and other small invertebrates. They can be completely aquatic (as the one in Figure 10.3, for example), or completely terrestrial, or they may live in the water for only part of their lives.

Figure 10.3 A salamander inside an aquarium. This salamander is an adult male, and it measures 6 centimeters in length, including its tail. It is called fire-belly newt because of the bright red-orange color of its belly.

FIGURE 10.3   This inscription exemplifies the relations between text and figure. You notice that there is a title, a text that contains a reference to the figure ("Figure 10.3"), a photograph, the figure number, and a caption. All these (title, text, reference, photograph, and caption) contain different information and are resources for your sense making.

detailed information about salamanders. At a certain point, the text presents an indexical reference ("Figure 10.3") that points you to a specific figure.

This figure is *illustrating* the topic of the text, that is, "aquatic salamanders." Conversely, the text *motivated* the selection of this particular photograph, that is, because the text is about salamanders, the photograph chosen is one that represents a salamander instead of, for example, a frog. The caption functions to *describe* what you can see in the photograph, that is, it describes the particular kind of aquatic salamander that is represented in the photograph. Notice that this caption presents more information than the caption on Figure 10.2; this extra information helps you in identifying what you can see in the photograph. The photograph *validates* the caption as long as it presents visual information that is equivalent to the textual information in the caption, that is, you are able to see in the photograph what you have read in the caption, although in this particular case, you are not able to see the color of the salamander's belly. The caption is actually providing extra information that the photograph is not validating, in the sense that you must believe that the salamander's belly is red-orange, but you cannot see it yourself in the black-and-white photograph.

Thus, to make maximum use of photographs (or other inscriptions), you need to trace the relations between title, text, caption, and inscription. You can also ask questions: how is the photograph related to the title? What does the text tell about the photograph? What is shown in the photograph about the topic that the text or caption does not tell? As a reader (or a teacher during classes, or a parent helping a child read the textbook), you must work through all the information provided, connecting these different resources in order to interpret them. To properly *read* the texts and the photograph, you must first understand how these resources are associated, and then you should be able to combine visual and textual information to arrive at an expected interpretation.

GRAPHS

In science, graphs are the most frequent of all the different inscription types. Any middle and high school textbook will contain a certain number of graphs—usually, the older the target audience, the more graphs can be found in the science textbook. Among the different graphs, we find pie charts, histograms, scatter graphs, line graphs, and many other types. However, many science textbooks fail to provide students with resources for learning how to use graphs for making sense and creating understanding. Let us take a brief look at the characteristics of graphs, how to make sense of them, and how to use them to create understanding.

Graphs that display the relationship between two variables may show a lot of detail, including specifics of the variables measured, the values of the measurement, the possible measurement error, and so on. An example of a graph including a lot of detail can be found on the very left in Figure 10.4; it

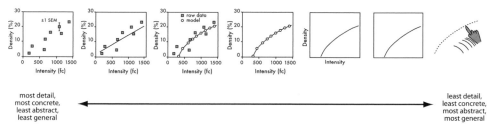

FIGURE 10.4   Line graphs can be ordered along a hierarchy from *most concrete*, containing the most detail, to *least concrete*, containing the least detail. This hierarchy can also be expressed as going from the most concrete but least general to the most abstract and most general.

depicts the relationship between light intensity (measured in foot candles [fc]) in some area and the density of some plant (in percent area covered). On the other extreme on Figure 10.4, we find graphs that contain very little detail, sometimes only a line, and in scientific conversations, such graphs are often not even drawn but only gestured in the air. That is, the graphs become more general as you go from left to right in Figure 10.4.

On the third graph from the left in Figure 10.4, you see circles added on a curvilinear line. You also find a *legend*, which tells you what the squares and circles stand for; the former stand for the actual measurements (the *raw* data), whereas the latter stand for some model. What the scientists are communicating with such a graph is that they have created a theoretical model, and in this model they have identified the theoretical relationship between light intensity and plant density. The model is something abstract, so in moving from the second to the third graph from the left in Figure 10.4, scientists have created and displayed graphs that are more abstract than the original data. As we move further to the right, there comes a point where the scientists leave the scales out (fourth graph from the left). Leaving out the scales means that readers no longer have a resource for making connections between the displayed graph and the measurements and measuring instruments they are familiar with; they can no longer bridge the gap from the graph to the real, lived-in world that is apparent in Figure 10.1. Without a specific scale, different sets of scales can be placed on the axes, which means that the graph has become more general, being able to represent not just the particular kind or type of plant but also plants in general. In the penultimate graph, even the labels have been omitted, so that now the graph is so general that it can be used to stand for quite different phenomena. The more abstract a graph, the more information the *reader* has to supply to make sense of it. Unfortunately, many textbook authors make use of graphs that are more like those to the right rather than those that are further to the left in Figure 10.4. Consider, for instance, the graph in Figure 10.5 as part of a chapter dealing with genetic selection and drift.

This type of graph requires you to go to the text and seek information about what is being displayed: you have to follow the same tracing strategy that we

If directional selection occurs, an extreme phenotype is favored over other phenotypes.

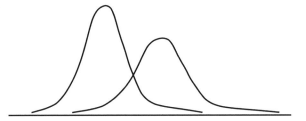

FIGURE 10.5   Graphs of this kind are often found in high school science textbooks. There is very little information in the graph that the reader can use to make sense; there is not even a vertical axis, a label, or a scale. Even the caption (above) does not tell the reader what the horizontal axis represents or whether the left or right curve represents the normal compared to which the other curve is "an extreme phenotype," as referred to in the text.

outlined for photographs. To begin with, it is a good strategy to actually write additional information into the book or to create a copy of the graph and write onto it any other information that might help you (or your student/child) in making sense of what the graph is about. For instance, you might want to think about possible variables that might have something to do with "phenotypes." For example, the height of some organism might be depicted on the horizontal scale, and the vertical scale refers to relative frequency. One or two hundred years ago, the average size of people was smaller (curve to the left) than it is today (curve to the right). Or the horizontal scale might be foot sizes: a recent report (*News from France* 2005) stated that over the past thirty years, the average foot size of the French has increased by 1 inch. Because the increase was the same for women and men, the overall distribution remained the same, which is just what the graph represents. We can now see that the distribution of the height (or foot size) of people in some population has not changed itself, just shifted to the right, which means, on the average, the entire population has increased its height or foot size by the *same* amount. To increase your understanding of the graph, you could ask questions such as "What created such a shift?" You may find this information in the textbook or in other sources; changes could be, for example, in the environment. Human beings eat more refined food and lead more protected and less dangerous lifestyles, and foods with certain properties are more widely available.

MAKING MORE OF INSCRIPTIONS AND TEXTBOOKS

Inscriptions are pervasive in today's society, and they will continue to be. Therefore, students, teachers, and parents need to pay attention to the possibilities of using inscriptions to understand and learn science. Inscriptions present indispensable occasions for students to learn science. Parents and teachers can help students to develop higher levels of visual literacy by actively working with them and linking them to other pieces of information in the book and also in real, lived situations at home and school. For instance, many textbooks present two different types of inscriptions as part of the same figure; these are called *layered inscriptions*. When a graph and a drawing are layered, for example, the more familiar inscription (the drawing, closer to the left side of the continuum in Figure 10.1) helps to bridge the gap between the least familiar inscription (the graph, closer to the right side of the continuum in Figure 10.1) and the real, lived-in world of the students. Teachers can help students also by projecting inscriptions onto a screen, and going through them with the students, asking relevant questions about what is represented in the inscription, how it was designed, and so on. Parents at home could do the same with their children when helping them studying science and doing their homework.

REFERENCE

News From France. 2005. Bigger French Feet. www.ambafrance-us.org/publi/nff/ NFF0502.pdf.

ADDITIONAL RESOURCES

Bertin, Jacques. 1983. *Semiology of Graphics: Diagrams, Networks, Maps.* Madison: University of Wisconsin Press.

Lynch, Michael, and Steve Woolgar, eds. 1990. *Representing as Scientific Practice.* Cambridge, MA: MIT Press.

Roth, Wolff-Michael, Lilian Pozzer-Ardenghi, and JaeYoung Han. 2005. *Critical Graphicacy: Understanding Visual Representation Practices in School Science.* Dordrecht, the Netherlands: Springer-Kluwer.

St. Martin, Fernande. 1990. *Semiotics of Visual Language.* Bloomington: Indiana University Press.

11

Science and Writing

Brian Hand

INTRODUCTION

Writing is not speech written down, that is, writing is a different act from talking. The importance and significance of writing in learning science are noted by the fact that in every science classroom, students will be asked to do some writing. Such writing may be as simple as writing down notes from the board, or become more complex in form such as taking laboratory notes or writing chapter summaries. With the current emphasis on science literacy, much debate hangs over the questions of whether students need to learn how to use language or need to use language to learn. These two positions place a completely different emphasis on what should occur in a classroom.

WRITING AS A LEARNING TOOL

Much has been written about what is occurring when we write a piece of text. There is current recognition that writing involves having to engage with two different knowledge bases—a content knowledge space (science knowledge) and a rhetorical knowledge space. When we write a piece of text (as I am attempting to do now), there is a need to deal with the content knowledge that is required for the subject matter, and an understanding of the rhetorical constraints (in this case, that the writing is the form of a chapter with particular constraints on style, amount of text, and audience). Engaging in writing that promotes the interaction of these two spaces is often referred to as *knowledge transformation*, because the knowledge being written about is being transformed from its stored stated, or is referred to as *knowledge constitution*, because the knowledge being written about is reconstituted into a new form. Writing that simply requires a memory dump is referred to as *knowledge telling*, that is, writing that only requires a recall of knowledge or basic writing of notes from the board.

To be beneficial, writing in science classrooms needs to be focused on the knowledge transforming or constituting area. We need to ask students to use writing such that they build their understandings of science—not to keep telling us what they already know.

LEARNING HOW TO USE LANGUAGE

Much of the current emphasis on the No Child Left Behind Act is to ensure that all children can read and write. All children are required to learn how to construct text, that is, build paragraphs, write complete sentences, use full stops and capitals, as well as be able to write using different writing types, for example, letters and reports. In the case of science, students are required to be able to understand how to write a laboratory report or a chapter summary. This means that students are required to know what a title, a purpose, procedures, results, and conclusions are so that they can successfully complete the task of writing a laboratory report. The argument is that getting students proficient in using science genres will lead to understanding of science concepts. Unfortunately, to date the research has not been clear about the benefits associated with this line of argument. Many of the people pushing this form of language instruction do not have backgrounds that are grounded in the sciences and do not understand some of the particular demands of the discipline.

While correctly arguing for the importance of having students become proficient in using language, the research to date is inconclusive as to the learning outcomes associated with this approach with respect to understanding science concepts. That is, there is limited evidence to argue for mechanical proficiency in the use of science writing strategies leading to gains in conceptual growth of students.

USING LANGUAGE TO LEARN

The focus of using language to learn is that learners need to be placed in situations where they have to use the language as a tool for building understanding rather than simply becoming mechanically proficient in its use. The orientation is knowledge transformation, not knowledge telling. It is by using language as a learning tool that students will have to become proficient in its mechanical use. As such, the emphasis is on both the science knowledge base and the rhetorical knowledge base. By having to write using different types of writing, to different audiences, and for different purposes, students will have to engage with the science content knowledge in a variety of situations and as a consequence build greater understanding of the topic. The type of writing that students engage in requires them to translate their understandings across a number of contexts, which promotes learning.

TRANSLATION ACROSS CONTEXTS

The writing that students do in classrooms is traditionally to the teacher only, that is, the teacher marks all work that is handed in. All the students who we have interviewed over a range of different studies have told us that when writing to the teacher, either they use "big words," which most times they don't really understand, or they expect the teacher to fill in the blanks because "they know what I mean." In other words, students hardly ever move beyond the terminology or definitions that the teacher presents because they believe that this is what the teacher wants (and one has to agree in principle with this line of reasoning). Students are rarely placed in a position where they are required to translate the science language of the classroom into everyday language for somebody else.

We have to remember that when sitting in class, students are trying to understand the science language being used, and the only frame of reference they have is their own language. Thus, it is important for us to push them to marry their everyday language with science language. By asking students to write for different audiences, we can begin to get them to move between science language (as discussed in the classroom) and the language used by their audience. For example, if we ask Year 8 students to write to Year 4 students explaining what they have learned about simple machines, then there is going to be a requirement for the Year 8 students to use language that the younger children can understand. Importantly, words such as *force, work, energy,* and so on need to be explained in a manner that young children can understand.

By engaging in such tasks, students are required to translate the language a number of times. First, they have to translate the science language into language that they themselves can understand; second, they have to translate this understanding into language that the young children can understand; and third, they have to translate back into science language when engaging in classroom activities, including tests. The research to date on the use of these strategies within science classrooms is indicating some very encouraging results in terms of improved student performances.

DIVERSIFIED WRITING TYPES

To promote the concept of translation across contexts, there is a need to diversify the types of writing that are traditionally done in science classrooms. Traditionally, students are involved in note taking, laboratory reports, chapter summaries, answering questions from the end of the chapter, and occasionally a poster project or creative story. However, as I said before, these types of writing are all marked by the teacher and tend to be about students focusing on pleasing the teacher by using the right words. There is a range of different writing types that we have been using in our research that appear to help students better understand science.

We have asked students to write letters to younger students, newspaper articles for the general public, brochures for their peers, PowerPoint presentations for their peers, and textbook explanations for younger students. Each of these types of writing has been intended for a different audience than the students traditionally write to. Importantly, all the audiences have been involved in assessing the written pieces, that is, when using diversified types of writing, we have always asked the intended audience to have input on whether they can understand science concepts described in the piece of writing and if the piece of writing is well written. This is critical to success when using writing as a learning tool.

ELEMENTS OF A WRITING TASK

Vaughan Prain and Brian Hand (1996) published a paper in which they described the various arguments about using writing to learn, as well as describing a model detailing the elements of a writing task. They indicated that there are five key elements of a writing task: topic, type, purpose, audience, and method of text production.

Topic

When writing, there is obviously a need to write about a topic. Given the emphasis on science standards and the focus on the "big ideas" of an area, the concept of topic refers to dealing with key concepts, linking themes within a topic, and writing about factual knowledge or the application of concepts. The students need to be able to write about how the key concepts of topic link to each other as a way to construct a rich conceptual framework. The critical element of topic is that when students are writing about a science topic, they are writing about science content knowledge and the product is not science fiction.

Type

Diversified writing types, as discussed above, relate to having students write such products as instructions, letters, concept maps, travelogues, and diagrams. Students are encouraged to write in ways that are different from the traditional forms of science writing. An important component of this element of type is the need for the students to ensure that they understand the rhetorical constraints/criteria necessary to fulfill the conditions of the particular writing type (e.g., a letter is not a travelogue).

Purpose

Writing can serve different functions depending upon whether the writing task takes place at the start of, during, or on completion of the unit. At the start of the unit, writing may serve the purpose of review, exploration, or devising a plan. During the unit, writing may serve the purpose to clarify, consider, or interpret; while on completion of the topic, writing may serve the purpose to demonstrate, revise, or test knowledge constructed throughout the unit.

Audience

Students can be asked to write for a broad range of audiences including peers, consumers, younger students, parents, and visitors. The important feature of this element is the need for the audience to be involved in the process. The audience is not a pretend audience, that is, the teacher should not set a task of writing to a different audience and then be the sole arbitrator of judging the writing. The reason is that the students will understand that the teacher is going to be the audience, and thus need be involved in the translation of the language as deeply as when they write to the audience intended.

Method of Text Production

Students can be asked to write a piece of text individually, but they also can be asked to work in pairs or groups. Given the current emphasis on technology, students can be asked not only to write using traditional pen and paper work, but also to use computers as a method of text production.

IMPLEMENTING WRITING TO LEARN STRATEGIES

Getting students to gain maximum benefit from using "writing to learn" strategies is not as simple as asking or requiring them to complete a writing task. There are many teaching strategies that need to be put in place. Recent studies have shown that there are a number of critical features that teachers need to engage with or promote to ensure that students maximize the benefits of completing such tasks. There are numerous "prewriting" activities that need to be completed.

Writing is about negotiating meaning—a writer is constantly trying to make meaning from the text he or she is constructing. Within a classroom, there are a number of opportunities that need to be provided to students to assist this negotiation process. Students need to be engaged in negotiation processes within whole-class, small-group, and individual settings. When a writing task is given, students need to talk through the rhetorical demands and subject matter associated with the task within small groups, have opportunities to present and defend their ideas before the whole class, and then be required to individually complete the writing assignment. Such opportunities allow each writer to orally discuss, debate, and defend his or her ideas as a means of negotiating the task before having to do it.

Students need to have an understanding of the rhetorical demands of the writing task. In science classrooms, we tend to think that our students will understand what is a letter, or a brochure, and so on. However, this does not mean that they understand the rhetorical demands that need to be completed: for example, in writing a letter there is a need for an introductory paragraph, some paragraphs that discuss the subject of the letter, and a concluding paragraph. Importantly, students need to understand that each paragraph dealing with subject matter must deal with one particular aspect, rather than spanning over a whole range of issues that are not connected. As not all science teachers are comfortable with this area of teaching (as often the comment is heard, "I am a teacher of science, not English"), this is an opportunity to involve an English teacher in helping students understand the demands.

As well as the rhetorical demands of the structure of the writing type to be used, there is a need for the rhetorical demands related to the purpose of the writing to be explained. For example, if the purpose of the writing is to argue for a particular position, then students need to understand the structure of an argument. Argument is not summarizing, and thus students need to understand that there are definite rhetorical demands for each particular purpose. Explanation, argument, summarizing, and so on are distinctly different, with separate rhetorical elements that need to be dealt with in the text produced by students.

A critical area for maximizing the benefits of using writing to learn strategies is the focus on conceptual ideas that frame a unit. Writing to learn strategies focus on assisting the construction of concepts, that is, knowledge transformation rather than knowledge telling. Thus, teachers need to align their teaching with requirements of the National Science Standards or relevant state standards, which require a focus on the big ideas of the topic, that is, teachers need to focus planning on the big ideas of the unit. Traditionally, science teachers have been concerned with ensuring that students know all the facts related to a topic, and expect the students to somehow build a conceptual understanding of the topic—as though this occurs by magic! Thus, teachers need to reframe the teaching unit around the big ideas of a topic and then, by using writing to learn strategies, engage the students in discussing/describing the concepts framing the unit.

When framing writing to learn strategies, teachers need to nominate a particular audience and ensure that the audience is real, that is, if the audience is younger students, then there is a need to organize the audience ahead of time. A critical element of this process is to frame a scoring matrix or rubric for the student authors and student readers. The writing task is asking students to deal with both science concepts and rhetorical elements, thus there is a need for a scoring matrix to pay attention to these areas. For example, such criteria could include the following: are the big ideas clearly explained, are the big ideas connected to each other, does the piece of writing flow, and are there just some criteria that can be used? The student authors need to understand that their readers will be sending a score back to them. This matrix should be given to the student authors at the start of a writing task in order to help them understand the demands of the task.

Teachers need to make decisions about how the text is to be produced. Given the emphasis of computers in school environments, teachers need to make decisions about what can be included within text. As described in other chapters, diagrams and graphs are just two different forms of representation that can be inserted into text. Importantly, discussions with students need to include how to ensure that the insertion of a graphic/diagram adds to and supports the text. This is a critical component in terms of building conceptual knowledge because the ability to represent knowledge across multiple forms indicates a rich understanding of the big ideas of the unit.

AN EXAMPLE

When coming to the end of a chapter on ecology, the teacher of Year 10 students may decide that he or she would like the students to write to a different audience about what they have learned. To do this, he or she needs to ensure that the major ideas framing the unit become the central point of the writing. Thus, the following steps need to be taken:

- Deciding on the big ideas—in this case, the idea that all ecosystems are balanced systems.
- Choosing the particular audience—peers, parents, or a younger audience—in this case, the audience is Year 8 students, and thus there is a need to work with a Year 8 teacher who will allow his or her students to read the artifacts produced.
- Choosing the writing type the students will undertake—the teacher decides that the students need to write a textbook explanation and prepares an explanation of what is included in a textbook explanation.
- Choosing the method of text production—in this case, the teacher plans to have the students do their planning in small groups but requires each student to complete his or her own piece of writing.
- To assist in the evaluation process, the teacher prepares a scoring matrix that includes both science content and rhetorical elements.

While this serves as a single example, the range of possibilities is endless and only limited by how much we as teachers want to push our students.

SUMMARY

Implementing writing to learn strategies within science classrooms is critical for building conceptual understanding of science topics. Writing is not about taking notes off the board or simple knowledge telling. The cognitive demands of implementing knowledge-transforming strategies are much greater and lead to better understanding by students. Thus students need to write using different types, for different purposes, and to different audiences. Teachers will need to change their ideas about the value of writing and the pedagogical strategies that they will need to use to gain value in using writing to learn strategies.

REFERENCE

Prain, Vaughan, and Brian Hand. 1996. Writing for Learning in Secondary Science: Rethinking Practices. *Teaching and Teacher Education* 12:609–26.

ADDITIONAL RESOURCES

Holliday, William G., Larry D. Yore, and Donna E. Alvermann. 1994. The Reading–Science Learning–Writing Connection: Breakthroughs, Barriers and Promises. *Journal of Research in Science Teaching* 31:877–93.

Klein, Perry. 1999. Reopening Inquiry into Cognitive Processes in Writing-to-Learn. *Educational Psychology Review* 11:203–70.

Yore, Larry D., Gay L. Bisanz, and Brian M. Hand. 2003. Examining the Literacy Component of Science Literacy: 25 Years of Language and Science Research. *International Journal of Science Education* 25:689–725.

Yore, Larry D., Brian M. Hand, and Marilyn Florence. 2004. Scientists' Views of Science, Models of Writing, and Science Writing Practices. *Journal of Research in Science Teaching* 41:338–69.

12

Reading and Science

William G. Holliday

Reading and science teaching are seldom discussed at the national level by science educators but remain big concerns for practitioners at the state and local levels. This chapter discusses students' reading and comprehension of printed science information, and provides some approaches for developing and possibly improving the effective integration of reading and science embedded in elementary schools. As students grow older, their reading problems seemingly worsen as they move into middle and high school science classrooms. Although researchers offer some promising advice on how to ameliorate these problems encountered by students in middle and high school, science teachers unfortunately are seldom provided with the needed administrative and instructional support.

READING IN ELEMENTARY SCHOOL

Learning How to Read Science

Students learning school science logically must learn how to read about scientific ideas and effective ways of tackling science-based problems. As these students mature, their ability to read about science and read other kinds of informational texts will become a great advantage as they cope with life's challenges. Students must learn in elementary school how to read by effortlessly decoding, recognizing, and understanding frequently encountered words, and by learning how to understand and remember what they read.

Students also need to become fluent readers, automatically recognizing words, developing reasonable background knowledge about their world, and effortlessly comprehending challenging texts.

Ultimately, students need to learn and apply with ease reading comprehension strategies so that they are equipped to inquire into the world of science. Specifically, science teachers need to help students learn about such strategies as how to identify main ideas, summarize reading selections, generate mental images of concepts, clarify fuzzy ideas, ask or generate questions about information located in the text, elaborate in writing on a key phrase located in a text, link unfamiliar scientific information with what they already know, work with their peers to tackle reading problems, and monitor their comprehension while reading science texts.

The fact is, students who cannot decode words cannot comprehend texts and cannot read fluently. These students who cannot comprehend texts with automaticity are unlikely to survive in the upcoming adult world, which is full of literacy challenges. The reading facts about modern life are that simple, in one sense. The world is full of informational texts in need of understanding and remembering.

No one knows exactly how to best teach reading, including comprehension reading strategies. Still, we know what needs to be done to improve students' reading comprehension of texts (located in places like textbooks, trade books, articles, and documents appearing on the Internet) such as those used in school science. Yet, we seem unable to excite leaders in science education at the national levels and teachers at local levels to take serious steps toward implementing what we know about how to teach reading comprehension strategies in science classrooms.

Few researchers in science teaching investigate how to teach reading comprehension to students, as evidenced by a review of research journals aimed at improving science teaching. In addition, few teachers apparently spend class time teaching students how to learn new vocabulary words located in science reading materials, and how to decide which logical reading strategies might be implemented to increase reading comprehension.

In elementary school grades, teachers spend lots of time teaching sound-letter, letter-word, and word-recognition skills and some comprehension strategies helpful when reading narratives like stories. But most teachers spend very little or no time teaching students how to comprehend informational texts used in science and social studies.

Younger students at good elementary schools engage in systematic reading instruction: they spend time learning about the alphabet, learn about relationships between symbolic letters and sounds, learn how groups of letters blend together to form words that have meanings, and understand that words blend together to form sentences with even greater complex meanings. Elementary school teachers do this in part through a combination of strategies—by teaching decoding by analogy and phonics approaches to learning new words, and by using some whole-language strategies linked to writing, for example.

Good readers analyze new words and bring together individual sounds in their quest to learn how to pronounce unfamiliar words, particularly in science. Readers also use parts of words—prefixes, suffixes, word roots, and combining forms—useful in decoding unfamiliar words. Many science textbook publishers in their upcoming editions surely will increase their instruction on using frequently encountered word chucks (e.g., prefixes, suffixes, root words, and combining forms) so that students can more easily learn new vocabulary words. These younger students are often taught using synthetic phonics programs or decoding-by-analogy approaches, relying on decoding approaches that potentially allow them to read varying texts including science texts with increased ease, accuracy, and efficiency.

Good phonics instruction in elementary school helps students understand that letters represent sounds, and that those sounds blend together to form words. As students become more familiar with phonics and with frequently encountered words, automatic word recognition results from a great deal of purposeful instruction followed by practice. That means that students later don't have to sound out large numbers of newly encountered words—continuously sounding out unfamiliar words. If less effort is devoted to figuring out what words sound like and what they mean, then students can devote more effort to understanding and remembering the message embedded in texts. So, students need to become familiar with a reasonable number of words so that they don't have to spent precious time, energy, and thought decoding unfamiliar words, a process that starves students' cognitive capacities—brain power, so to speak—from the more immediate job of comprehending authors' printed messages.

Who is teaching students how to comprehend informational texts such as printed materials used to learn science? One problem in reading is that elementary school teachers mainly focus students' attention on how to comprehend narratives such as stories about familiar topics, including selections from American and English literature.

Unfortunately, today's young people (and perhaps many readers of this chapter) more typically learned how to comprehend informational text using unsystematic, trial-and-error methods (Nokes and Dole 2004), fending for themselves when it came to learning how to comprehend the kind of printed text most often encountered in their schooling and on-the-job training, and as adults trying to keep up to date with important advancements in science and other information-packed domains. One approach to grappling with the problem of teaching science and reading has been explored by three separate groups of teachers and researchers working with fourth- and fifth-grade students. Common factors identified in these successful programs are explained in the next section.

BUILDING AN INTEGRATED SCIENCE AND
READING PROGRAM

Science teachers working with other upper-level elementary school teachers might consider developing their own integrated science-reading program—science and reading teachers working together. This approach reportedly has proven useful to teachers developing their own programs—integrated programs developed by teachers and documented by researchers.

Integrating reading with science makes political and research-based sense, especially during these times when reading and math are dominating so much of today's elementary school curriculum. Integrating reading with science also may provide a greater opportunity for teachers to pursue their professional goal of delivering a superior science program to their students. Let's examine some of the features of three successful science-reading programs.

First, these programs seem to take a few years of teacher effort, persistence, patience, cooperative administrators and parents, willingness for everyone to take some risks, and some external stimulus to spark and excite teachers into working productively together. Teachers often become energized when teaching choices are left to their discretion rather than dictated from administrators or researchers. Some political compromise, of course, is always necessary when embarking on a new project of this kind.

Teachers developing these programs focused on motivating students—working toward getting students to want to learn, solve problems, tackle difficult challenging issues, and work cooperatively. This combination theoretically is a formula for increased academic success and learner independence.

Second, the evidence from selected multiyear studies suggests that integrating writing into each of these integrated programs helps students read better and learn how to inquire into their physical world. Students need to express themselves verbally in writing, which seems to clarify their understanding of what they read, a common element in whole-language approaches to literacy education. Still, teachers need to provide explicit instruction on how to inquire and read, model how they as teachers might tackle problems and support their students, reduce over time the support provided to their students, and get students working toward greater independence.

Third, students in successful programs often develop cooperative group projects and are provided authentic, nontrivial, learning choices. Such projects can be helpful because they motivate students to share ideas, use search-reading strategies, plan manageable learning goals, engage in problem-solving events, and tie their background knowledge to their newly discovered science knowledge. Of course, academic life even for elementary school students requires effort and is more than engaging in exciting projects. Students need to learn that learning is not always fun, and need to learn how to work hard, study independently, and achieve in science and reading.

Fourth, successful students often engage in hands-on activities, tying what they explored in a concrete sense with their more abstract learning by reading trade books and other materials. An apparent key to success is not spending an extraordinary amount of time just doing hands-on activities. Instead, additional teacher-guided activities—involving science inquiry processes like observing, experimenting, and inferencing—provides a foundation from which abstract information less familiar to students is brought to bear on science-based issues and problems. Such activities apparently can motivate students to ask questions seldom seen in traditional science and reading programs. Motivational studies suggest that many struggling students can overcome academic handicaps and begin to display behaviors resembling students who appear bright, know a lot, come from "great" homes, and mingle with informed people.

Fifth, these students learn about reading comprehension, monitoring their learning progress, setting their goals, making productive choices, asking questions, discussing observations, and reading and writing on topics with their peers. These students also need systematic explicit instruction in reading, including phonemic awareness and comprehension instruction dealing with informational texts like those found in social studies, math, and science classes. These teachers working with researchers learned that developing learning strategies in children took many years of concentrated effort to cultivate and develop, often introducing a few reading strategies at one time.

READING IN SECONDARY SCHOOL

Attacking Reading Comprehension Problems

Students at all levels of schooling who cannot comprehend printed text when presented with challenging ideas about science are in deep trouble—academically, vocationally, and socially.

The Problem

In secondary school grades, a few science teachers wrongly assume that many students can cope with a wide range of words and can comprehend informational texts presented in school textbooks and other materials. Most other teachers are realistic and skeptical of students' reading comprehension abilities.

The evidence suggests that about 60 percent of middle school students can basically read but most have trouble comprehending and remembering what they read, according to data reported by the National Assessment of Education Progress' (NAEP 2003) *Nation's Report Card*, and an analysis by the Alliance for Excellent Education (2004). They concluded that most students are not "proficient" readers, *proficient* meaning " 'solid academic performance' for the assessed grade" (Alliance for Excellent Education 2004, 7). The good news is that many of these students can recognize many words and read at a so-called basic level—"partial mastery of prerequisite knowledge and skills."

That is, these students learned how to read simple prose during their elementary school years, but only about one-third of our eighth-grade students can read and comprehend informational texts with proficiency, optimistically speaking. Students at the basic level cannot relate the printed information in front of them with their prior knowledge—necessary background knowledge—needed to understand the topic being read. Moreover, these struggling readers cannot transfer what they read to a slightly new situation or context by answering simple application-style questions—questions beyond recall, rote memory-style questions. Even many college students reading textbooks have considerable trouble comprehending the material.

The further bad news is that science teachers in secondary schools are not devoting class time to teaching students how to comprehend information from their textbooks and other reading materials. Language arts teachers including reading teachers need the help of science teachers because students need to learn how to comprehend science textbooks while under the supervision of a teacher who fully comprehends the science being taught and other critical attributes of good science teaching. Students are typically left to their own devices and forced to figure out ways of understanding the science presented in texts. These students need reading strategy instruction that works, practice in implementing effective strategies, and knowledge about when, why, and how to apply reading strategies while learning from printed science materials. For example, students unfamiliar with reasons why reading strategies are important often fail to implement them during studying.

Attacking the Problem

We need to improve both classroom strategy instruction and supportive school infrastructure. In other words, research in reading education focusing on secondary school literacy problems in areas like science teaching recommends bolstering reading comprehension instruction embedded in a supportive school environment, according to a recent influential report produced by a group of scholars (Alliance for Excellent Education 2004). One of the main findings from their study was that most students are ill prepared to comprehend content-based reading in domains like science. This blue-ribbon committee concluded that requesting willing teachers to implement a set of instructional strategies was likely to fail if the infrastructure of the school was not altered to support such implementations. This conclusion was framed in two separate but interdependent groups of recommendations.

First, instructional improvements should

- use explicit comprehension strategy instruction where teachers explain and model how good readers read a challenging text rather than just relying on students to learn such strategies on their own and hoping that students acquire such strategies through indirect or so-called natural teaching methods;

- implement ways of embedding instruction into existing subject-area courses where science teachers, for example, actually teach students how to comprehend;
- capitalize on achievement-motivation research by emphasizing the need for normal students, for instance, to work hard, be persistent, and refuse to accept flawed biological "innate" reasons for their inabilities to comprehend challenging texts;
- create and maintain cooperative, small-group instruction wherever practical to facilitate improved reading because peers working together to support each other coupled with some teacher support have proven to comprise a successful way to teach reading strategies;
- institute a remedial instructional program for struggling students unable to decode and lacking reasonable prior knowledge, motivational attributes, word-recognition facility, and basic comprehension fluency, rather than asking content-area teachers to help such students or hoping that more time will "naturally" turn these students around;
- provide practice for students consisting of reading a wide range of informational texts beyond just narratives, stories, and nonchallenging printed materials;
- link writing opportunities with what students are reading so that students learn the importance of planning, drafting, and revising lengthy written products based in nonnarrative texts; and
- collect data regularly on how students are progressing and identify specific problems so that corrective measures are woven into future instructional efforts to facilitate students' reading comprehension.

Second, these instructional interventions are of little use, concluded this blue-ribbon committee of scholars, if the infrastructure of schools is not modified in reasonable ways, including the following:

- extend the time in science classes devoted to reading instruction,
- collect data to assess student progress useful in understanding individual student problems,
- coordinate instruction by having teams of teachers regularly working together on a daily basis,
- insist on committed school principals and other administrators who will provide the funding and coordinated support needed for teachers to do their jobs, and
- construct a comprehension strategy instruction program among content-area courses like science and other school personnel.

This panel of scholars insisted that programs failing to implement even some of these and other recommended have little promise of success. The last two sections discuss two other problems. First, we need to find a way of convincing science teachers to commit to a reading instruction embedded in their science class. Finally, I will present some concluding thoughts.

FOCUSING ON SCIENCE TEACHERS

One of our biggest problems—say many reading educators—is convincing middle and high school science teachers to spend a little time teaching reading comprehension strategies to students who are struggling with science texts. Even "good" readers can profit from science teachers' help. Science teachers in secondary schools, of course, cannot teach students how to read at the basic level because extensive training is needed to teach students in middle and high schools how to decode words, teach them how to make sense of symbolic letters, and provide other forms of remedial instruction to nonreaders. Besides, such instruction takes years of tutoring by trained professionals with specializations on how to handle such severe literacy problems.

Still, why don't teachers typically devote some time, even a little time, to teaching comprehension strategy instruction in their science classes? No one knows for sure. This remains a big problem seldom mentioned and not taken seriously enough. But, if we analyze the problem, perhaps we can begin to produce some hypotheses leading to meaningful ways of introducing reading comprehension strategy instruction. So, here are some possible reasons for not teaching comprehension strategies that I have heard from practicing science teachers. Each one of them in one sense is rational, is understandable, and reflects the challenges of teaching science in today's schools.

1. The amount of science content that must be taught in today's classes precludes additional instructional time redirected to teaching reading strategies.

2. There is inadequate research-based evidence supporting the notion that any of the touted research-based, reading strategy instruction programs work under typical class conditions. We only have unconvincing anecdotal evidence and some hypotheses in need of additional data-based evidence collected under realistic science teaching conditions.

3. Teachers need to spend precious time engaging students in inquiry-oriented, hands-on activities including project-based efforts, which must supersede serious reading instructional considerations.

4. Teachers are strongly encouraged to prepare students for year-end, high-stakes, standards-based testing assessments, leaving little time to engage students in such strategy instruction.

5. The *National Science Education Standards* (National Research Council 1996) say virtually nothing about teachers devoting time to teaching students how to (specifically) read and comprehend information presented in students' textbooks or any other reading materials.

6. Teachers apparently relish opportunities to explain and clarify the meanings of science concepts through lectures and discussions with students rather than assume with good reason that students can learn portions of the material from

their science textbooks. (The available evidence suggests that most teachers are pretty good at explaining science concepts to their students.)

7. Students typically encourage teachers to explain in class science content in contrast to encouraging teachers to assign portions of challenging textbooks or other readings.

8. Teaching reading comprehension strategies might not be received well by students, perhaps resulting in increasing the time and energy needed to deal with abhorrent classroom management problems.

IN CONCLUSION

The *National Science Education Standards* (National Research Council 1996), perhaps the most often cited reference at the national level in science education during the past decade, directly says almost nothing about the importance of reading in science education. Perhaps the most direct comment, taken from page 22, states, "Scientific literacy entails being able to read with understanding articles about science in the popular press." Yet, professional scientists and scientifically literate laypeople apparently acquire much of their scientific knowledge through reading. These informed people heavily depend on their abilities to read and implement comprehension reading strategies needed to understand a variety of informational texts describing many topics, including scientific concepts and problem-solving strategies. But, you would never deduce this point by reading the *Standards* book and listening to science educators.

Science education researchers and practicing teachers at all levels of schooling have their work cut out for them if they intend to tackle this reading problem evidenced particularly in more than one-third of our eighth-grade students. We must prove to these devoted professionals that a relevant and pressing problem exists. But how?

How can we get science educators seriously committed in tackling problems dealing with reading strategies instruction?

I encourage people interested in finding out more about research-based findings linked to practice to read the following three books, in the order presented below:

Michael Pressley's 2002 popular book, *Reading Instruction That Works*, which focuses on basic reading and related motivational problems identifiable in K–12 schools and problems of reading comprehension

Well-received chapters in *Adolescent Literacy Research and Practice*, edited by Jetton and Dole (2004), which especially focus on reading problems found among students enrolled in secondary schools

Well-received chapters produced by the National Science Teachers Association (NSTA) and the International Reading Association (IRA) in *Crossing Borders in Literacy and Science Instruction*, edited by Wendy Saul (2004), which focuses on scholarly viewpoints about literacy and science teaching

REFERENCES

Alliance for Excellent Education. 2004. *Reading Next: A Vision for Action and Research in Middle and High School Literacy.* Washington, DC: Alliance for Excellent Education. www.all4ed.org.

Jetton, Tamara L. and Janice A. Dole, eds. 2004. *Adolescent Literacy Research and Practice.* New York: Guildford Press.

National Research Council. 1996. *National Science Education Standards.* Washington, DC: National Academy Press.

Nokes, Jeffery D., and Janice A. Dole. 2004. Helping Adolescent Readers Through Explicit Instruction. In *Adolescent Literacy Research and Practice,* edited by Tamara L. Jetton and Janice A. Dole, 162–92. New York: Guildford Press.

Pressley, Michael 2002. *Reading Instruction That Works: The Case for Balanced Teaching.* New York: Guildford Press.

Saul, Wendy, ed. 2004. *Crossing Borders in Literacy and Science Instruction: Perspectives of Theory and Practice.* Arlington, VA, and Newark, DE: National Science Teachers Association and International Reading Association.

13

Making Science Content Comprehensible for English Language Learners

Beth Wassell

Imagine walking into a high school chemistry classroom—in Mexico. You are already somewhat intimidated by the subject matter, as you consider yourself only a mediocre science student. Although you know some conversational Spanish, this is the first time you've ever been in a class where the teacher did not speak any English. Luckily, you have a few friends in the class that speak English fluently; however, the majority of the students speak only Spanish and appear to be offended when you and your friends address each other in your native tongue. The teacher begins her lecture and within five minutes, you are completely overwhelmed. Although you recognize some cognates such as *ciencias, laboratorio, experimentos, observación*, and *resultados*, you have a hard time following the instructor and deciphering the main ideas of her discussion. After a few more minutes of total frustration, you begin to tune out. . . .

WHO ARE ENGLISH LANGUAGE LEARNERS?

Although you may have never experienced something similar to the situation above, many students new to the United States have. As our country becomes more ethnically and linguistically diverse, educators must address the needs of *English language learners* (ELLs)—students whose primary language is not English. According to the National Center for Education Statistics (U.S. Department of Education 2003), as of the 2001–2002 school year, about 5 percent of public school students in the United States were identified as Limited English Proficient (LEP). The majority of these students are concentrated in urban areas, Western states, and larger schools with populations of more than 750 students. Between one-quarter and one-third of students in California have limited English proficiency. Approximately 42 percent of teachers in the United States report having at least one ELL in their classes (U.S. Department of Education 2003).

As we become more aware of issues that increase the achievement gap between white students and students of color in our public schools, it is vital that we consider the ways in which ELLs are falling behind, especially in academic courses like science. In this chapter, I describe some of the issues that confront English language learners in the context of science content classes. I then illustrate how sheltered instruction can help science teachers make content more accessible and comprehensible to ELLs, which will help them to become more scientifically literate. Included in the chapter are practical ideas that can be used to modify inquiry-based lesson plans and assessments for ELLs. Finally, I argue that science educators need to have increased awareness of ELLs' needs in order to foster their scientific literacy.

ISSUES CONFRONTING ELLS IN SCIENCE

Most ELLs receive specialized services, such as ESL (English as a Second Language) classes or bilingual education programs. According to Sonia Nieto in her book *Affirming Diversity* (2004), the policies and practices that surround ELLs in public schools vary, ranging from strong bilingual programs to "sink or swim" approaches, where ELLs are in English-only classrooms all day. Many schools have ESL programs[1] in which students spend one or more periods per day working with a certified ESL teacher with the goal of increasing English proficiency. For the rest of the day, students are usually enrolled in English-only content-area courses. However, students often flounder in such classes, especially when instructors use predominantly traditional, teacher-centered methods.

In regular, English-only science courses, ELLs confront multiple challenges. First, they are challenged with academic language that is especially difficult if they are unable to use English in functional or more conversational situations. Second, some students may use functional English comfortably and appear to teachers to be fluent, for instance, by proficiently interacting in basic, interpersonal, cognitively undemanding tasks, such as following directions or carrying on a face-to-face conversation about a homework assignment. Such situations could be considered context-rich, that is, the language used is highly embedded in context or takes place in typical communication situations. However, some ELLs may be unable to comprehend in situations where the language is cognitively demanding and separated from the context, such as in an explanation of an abstract physics concept (without diagrams) or standardized tests, which rarely include context clues (Cummins 2000). Thus, teachers are often confounded by students who seem fluent in English yet perform poorly on assessments.

Other issues arise when ELLs are exposed to less rigorous science content and curricula. ELLs may be placed in lower-level science courses and miss out on advanced courses that will more aptly prepare them for college. In other cases, teachers water down the content for English language learners, even though students may have been exposed to advanced science courses in their country of origin. In either case, ELLs are denied the experiences of challenging, high-level science courses and receive less than an equitable education.

Compounding the issues described above is the limited knowledge that content-area teachers have about strategies they can use to better serve ELLs; most teacher education programs do not include training in this area. I illustrate several tools that science educators can use to help ELLs achieve in science classes: sheltered instruction, language resources, cogenerative dialogues, and research.

MAKING SCIENCE CONTENT COMPREHENSIBLE

Sheltered Instruction

Sheltered instruction[2] is an approach to teaching content in a comprehensible way that also fosters students' English language development. Teachers' use of sheltering techniques can increase student success in science and enhance scientific literacy. In *Making Content Comprehensible for English Learners: The SIOP Model*, Jana Echevarria, Mary Ellen Vogt, and Deborah Short (2004) propose a sheltered instruction model to help teachers make content more comprehensible while teaching language skills. Although the model includes eight components for planning, instruction, and assessment, I will discuss two of the central elements: lesson preparation and review/assessment.

In *lesson preparation*, teachers develop not only content objectives but also language objectives that are incorporated into the lesson. The language objective should address the language skills students will need to understand the concept or perform an activity. For instance, in a lesson on buoyancy, the content and language objectives might be as follows:

> *Content objectives*: Students will be able to predict and calculate whether an object has buoyancy. Students will graph their findings.
>
> *Language objectives*: Students will *write* predictions about why something would float or sink. Students will *discuss* their predictions in groups. Students will *use* the words *float*, *sink*, *calculate*, *predict*, and *buoyant* appropriately in context.

During the planning phase, it is important to consider ways of tapping into the students' background and cultures. For the buoyancy lesson, teachers could ask whether students have ever tried to float in a pool or in a body of water, or what things generally float in a bathtub. Then, by using actual objects (e.g., an orange, a toy boat, or a small weight) and by drawing on their personal experiences, students could write predictions on whether they think the items will float or sink. It is important to keep a list of important vocabulary in the room that students can see and refer to throughout the activity (a large piece of chart paper or a bulletin board can be used). For the buoyancy lesson, key vocabulary would include the words mentioned in the language objectives.

Another component of the model is *review/assessment*, which addresses the review and assessment of both content and language objectives. Teachers should assess students' fluency with key vocabulary as well as the major content concepts covered. In addition, teachers should give students feedback on their output, or their oral and written responses. Restating or paraphrasing students' responses correctly, especially if their responses include glaring grammatical errors, can be a low-anxiety way of offering students feedback on their output. ELLs should always feel supported and validated when receiving feedback; negative feedback can raise students' anxiety and discourage them from participating orally in class.

When thinking about formal assessment in the science classroom, it is important that ELLs are exposed to various types of evaluation. For beginning ELLs, assessments that are highly contextualized are necessary. Tests and quizzes can be made more contextualized if they include diagrams, visuals, and other context clues. Alternative and authentic forms of assessment, such as portfolios and projects, should be used in addition to traditional tests. For instance, students can be assessed by making a prediction or hypothesis, demonstrating the procedure in an experiment, and discussing the results. Also, participation should always be included in students' grades.

COGENERATIVE DIALOGUE

Another way that teachers can help ELLs in science is by giving them opportunities to discuss their learning outside of the classroom in cogenerative dialogues. Cogenerative dialogues, which are discussed in detail in Sarah-Kate LaVan's chapter (see Chapter 31), are discussions among stakeholders (e.g., the teacher, students, and administrators) that focus on collectively generating outcomes that will improve teaching and learning in the class. In these small sessions that can be held after school, during lunch, or even during the class period, teachers and students can discuss their perceptions of the class. Cogenerative dialogues can focus on the way the material is taught, on the material itself, and on ways to encourage interaction in English. For instance, the cogenerative dialogue might focus on a lesson that has recently been enacted; together, students and the teacher can talk about the methods the teacher used to present the material and alternate ways the lesson could have been taught. It is important for students to feel as though their voices are valued, which means that teachers should strive to implement the ideas generated in the cogenerative dialogue. However, students and teachers should remember that the onus for change falls not only on the teacher; students, too, should have responsibility for enacting change. Co-generative dialogues that focus on pedagogy can potentially improve student understanding in subsequent lessons.

Cogenerative dialogues not only encourage feedback about the class but also can be used to assess student understanding of science concepts. The dialogue can be an informal forum to discuss specific concepts that have been covered in class. Teachers can then assess understanding by engaging students in conversations about the concepts. Together, students and teachers can decide which concepts need to be reinforced in future classes and how this could be done efficiently. Thus, teachers can directly influence curricular planning and incorporate alternative forms of assessment through cogenerative dialogue.

An important by-product of cogenerative dialogue is that it increases out-of-class time in which students are speaking English. Engaging ELLs in interaction through cogenerative dialogue can positively impact their language proficiency. Interaction that encourages elaboration on the lesson's concepts should take place between the teacher and students and among students. Allowing sufficient wait time for students to think and respond can foster such interaction—this is especially important for ELLs, since they often need extended time to process and formulate ideas in English.

USING LANGUAGE AS A RESOURCE

Another way teachers can help students better understand science content is by allowing them to use their native languages as a resource to clarify difficult concepts. Teachers can help foster such opportunities by pairing or grouping students when possible in "same-language" clusters. Simply having bilingual dictionaries available in the classroom can be extremely helpful when students encounter unknown words and phrases. Also, teachers can find subject-specific resources in other languages—many textbook publishers in the United States offer materials in Spanish. For writing assignments, beginning ELLs can be asked to first formulate their answers in their native language and then translate them into English.

Grouping is another important aspect that can foster interaction in the science classroom. If a variety of languages are represented in the classroom, groups that unite students with different first languages may further increase interaction in English. In sheltered instruction science classrooms, it is important to use a variety of grouping techniques so students have opportunities to work with other ELLs and native English speakers. ELLs should not be grouped together regularly out of convenience; they should have extensive opportunities to work cooperatively with *all* students in the classroom.

Teachers can also use language as a resource for ELLs to ensure that students comprehend the language of instruction and directions for learning activities. To do this, teachers must use comprehensible input, or speech that is clear, explicit, and repeated often. For instance, the steps involved in the buoyancy activity above could be presented orally to students, written on the board or overhead projector, and demonstrated. Throughout the lesson, teachers should speak slowly and enunciate words clearly. The use of gestures is also important—in the buoyancy lesson, teachers can use simple gestures to signify key vocabulary such as *float* and *sink* or to demonstrate the procedures involved in lab or hands-on activities. Input can be further clarified for students by the use of visual aids. In addition to the list of significant vocabulary for the lesson or unit, teachers can create large flashcards with pictures and their meanings in English. These can then be hung up around the room for students to use as resources.

RESEARCH IN THE SCIENCE CLASSROOM

As science teachers endeavor to use new practices in the classroom, they need opportunities to study their own teaching and its efficacy in helping ELLs become more scientifically literate. This can be achieved when teachers research their own practices in the classroom. Through action research, which, according to Andrew P. Johnson, is "a systematic inquiry into one's own practice," educators can make informed decisions in the classroom, evaluate teaching methods, and assess learning outcomes (Johnson 2004, xi). For example, teachers interested in helping ELLs achieve might consider one of the following questions:

- How comprehensible is my input (or instructional speech) for ELLs?
- What types of activities do I use that enable ELLs to grasp content and practice their language skills?
- How does my consistent use of key vocabulary lists/visual aids/hands-on activities increase my students' achievement levels in my class?

After establishing research questions, teachers can collect data in their classrooms, analyze it, and formulate findings that will directly impact practice. Teachers interested in exploring the first question above (regarding comprehensible input) could use a video camera to record several lessons and analyze the rate and clarity of their speech. Student interviews, surveys, or cogenerative dialogues could also be used to gain other perspectives on the questions from the learners themselves. See Chapter 26 to read more about high-quality teacher research.

CONCLUSION

Using sheltered instruction, cogenerative dialogue, language resources, and teacher research in science classrooms will not ensure the success of all ELLs. However, the ideas here offer ways that teachers can be proactive when planning for instruction, enacting lessons, assessing students, and rethinking their own practices. In order to rightfully address the issues facing ELLs, educators must consider specific ideological and policy concerns. I argue for three important changes in science education: awareness of ELLs' needs, increased and more effective professional development, and expanded roles for science teachers.

First, teacher educators, policy makers, and administrators must learn more about the immigrant student population and advocate for their specialized needs. Because the number of ELLs is growing exponentially in the United States, school leaders need to raise awareness of ways in which ELLs have been marginalized and positioned as "the other" as an important first step. According to Sonia Nieto (2004), many educators view ELLs with a deficit lens and claim that students "do not have language" or "lack language skills." Although these students may be at the early stages of acquiring English, they are not lacking language entirely—many ELLs are functionally proficient in their native language. Furthermore, teachers must consider a student's native language as an important part of his or her identity—an aspect that should not be dismissed. Since language is highly embedded in identity, students' native languages must be valued and, in optimal cases, developed further *while* learning English—students should never be asked to leave their language at the classroom door.

Teachers must also become advocates for ELLs' success. The parents of ELLs may work long hours, have little time to engage in school-based activities or meetings, and be unaware of the practices that middle-class parents do to help their students succeed in the academic arena. Parental roles in other countries are not necessarily congruent to those teachers often expect in the United States; thus, students should not be penalized as a result of this dissonance.

Second, both preservice and in-service teachers need opportunities to learn about methods that will help ELLs better understand science and other content-area concepts while developing English language skills. In some schools, ESL teachers have the sole responsibility for helping ELLs; content-area teachers have depended on them to modify assessments and to reteach content lessons that students did not understand in class. Some content-area teachers also claim that they are only science, math, social studies, or even English teachers and do not know much about teaching second language skills. For ELLs to be successful in inclusive classrooms, content-area teachers need to be cognizant of ELLs' needs and learn to tailor instruction for their benefit. This can only be realized through meaningful courses in teacher education programs and intensive professional development opportunities.

Finally, as the field of science education strives to reframe science teaching with the goal of making students more scientifically literate, we must acknowledge that this is an opportune time to expand the roles of science teachers. Although we cannot expect everyone to become a certified ESL teacher, we can hope that science educators will use sheltered instruction strategies that not only will help ELLs to better understand abstract science concepts, but also will help all students become more scientifically literate. With increasing numbers of ELLs in our nation's schools and mounting accountability measures prompted by No Child Left Behind, we can no longer ignore the ways the field of science education must be proactive.

NOTES

1. Also referred to as ESOL—English for speakers of other languages.
2. This approach is also referred to as SDAIE (specially designed academic instruction in English).

REFERENCES

Cummins, James. 2000. *Language, Power and Pedagogy: Bilingual Children Caught in the Crossfire.* Clevedon: Multilingual Matters.
Echevarria, Jana, Mary Ellen Vogt, and Deborah Short. 2004. *Making Content Comprehensible for English Language Learners: The SIOP Model.* Boston: Pearson, Allyn & Bacon.
Johnson, Andrew P. 2004. *A Short Guide to Action Research.* Boston: Pearson, Allyn & Bacon.
Nieto, Sonia. 2004. *Affirming Diversity: The Sociopolitical Context of Multicultural Education.* Boston: Pearson, Allyn & Bacon.
U.S. Department of Education, National Center for Education Statistics. 2003. *Overview of Public Elementary and Secondary Schools and Districts: School Year 2001–2002.* NCES 2003 411. Washington, DC: U.S. Department of Education.

WEBSITE

The California Science Project has resources for standards-based science instruction for ELLs, and includes several downloadable documents: http://csmp.ucop .edu/csp/initiative.html

14

The Interrelationship between Technological Fluency and Scientific Literacy

Lisa M. Bouillion and Donna DeGennaro

New technologies continue to influence the way we work, learn, communicate, and play. *Technology fluency* is increasingly argued to be a necessary skill for equitable access and participation in today's society. According to the National Academy of Engineering (2002), the benefits include improved decision making, increased citizen participation, a strengthened modern workforce, enhanced social well-being, and a narrowed digital divide. A recent International Technology Education Association (ITEA)/Gallop Poll (International Technology Education Association 2004) on how Americans think about technology reports that the public is virtually unanimous in believing that the study of technology should be included in the school curriculum. This expectation is supported by current education standards and is reflected in the recently released U.S. Department of Education National Technology Education Plan (U.S. Department of Education 2004).

Technology fluency is an integral component of what it means to be scientifically literate. That is, to generate and apply scientific knowledge, identify questions, and draw evidence-based conclusions one must be able to access and manipulate emerging technologies. Just as microscope technology has allowed us to view objects that are too small to be seen easily by the naked eye, virtual reality and other visualization programs are providing new opportunities for representing and interacting with scientific data sets. Likewise, Internet-based technologies are increasingly used to share these data and facilitate dialog around scientific findings. An important part of scientific literacy is the capacity to use these new technological tools.

EVOLVING DEFINITIONS OF TECHNOLOGY
FLUENCY IN EDUCATION

Perspectives on the role of technology in education have varied. Many initial integration efforts have placed technology learning outside curricular activities in order to support learning *about* technology. In this view, educators often seek to teach students discrete technical skills that might include simple tasks such as clicking the mouse, turning on the computer, or saving to a disk. Processes that are more complex might include searching the Internet and creating a document with a word-processing software such as Microsoft Word. While these discrete skills are arguably important, the technology fluency needed to support and engender science learning requires opportunities for students to learn and practice science *with* technology. In these cases, students draw upon previous technology skills while at the same time they might be required to adapt technologies to different ends, purposefully as well as dynamically, and interact with new technologies as questions and understandings of contextual learning evolve. Youth then develop fluency during the process of creating a product with and/or through technology. In this view, students acquire technology skills through goal-driven activities that provide opportunities for exploration, experimentation, and expression. Ultimately, the measure of technology fluency involves being an active *creator* with technology, in addition to being a critical *consumer*.

SCIENTIFIC INQUIRY AND THE CHALLENGE
OF NAVIGATING AVAILABLE DATA

For nearly three decades, scientists have been using some form of Internet communications to share resources, knowledge, and data with other scientists. More recently, the Internet has become a transparent tool, which in turn has profoundly reshaped the availability of information to a general audience. Moreover, the Internet has opened opportunities to alter practices of data collection and sharing among and across different kinds of users. A student interested in learning more about tsunamis, for instance, has only to type this single word into an Internet search engine (e.g., Google and Yahoo!), and in a matter of seconds she will receive a listing of over 22 million "hits," or links to related information. That information might come in the form of a layperson's interpretations of this phenomenon, an individual's firsthand account of experiencing the incident, or a scientist's real-time data of the event. The sheer volume of data available to students in their efforts to construct scientific explanations is staggering. The capability of critically filtering and organizing collected data for use in scientific experiments is an example of how technology fluency and scientific literacy are intertwined.

Several technological tools have been developed to assist students in filtering and organizing data collected from the Internet. For example, several search engines have been specifically designed to identify websites matching set criteria of appropriate content and reading level. Examples of such search engines include Yahooligans, Ask Jeeves for Kids, KidsClick!, and Think-Quest Library. Artemis, another example of an education-specific search engine, provides the additional support of helping students to keep in mind the question that drives their online data search. Following a prompt from the system, students record their initial question before conducting a search for information. Upon entering a question, a folder with the name of the question appears in their workspace. As the online inquiry evolves, students can pose new questions. For each new question, a new "question folder" is listed. Students record analyses of the collected information within each question folder. This online search tool also makes it easy for students to share their findings with others, and teachers are able to monitor the entire system in real time.

WebQuests and Web-based Inquiry Science Environment (WISE) are examples of technological tools that provide organizational supports for online investigations within the context of previously designed science inquiries. WebQuests are teacher-designed inquiry activities that draw at least in part from preidentified online data sources. A WebQuest includes a task objective, an outline of the inquiry process, and the expected outcomes. In some cases, the online inquiries are limited to these resources; and in other cases, students are encouraged to use these sources as a starting place for further inquiry. Resources are available online to assist teachers in developing their own WebQuest, but there are thousands of already designed WebQuests available for public use. For example, a search on Google for "biology web-quest" brings up 26,100 links to online inquiry activities ranging in topics from DNA fingerprinting to biomes.

WISE is a free online learning environment for students in grades 5–12. Currently, WISE offers over fifty science experiences. Some topics include genetically modified foods, earthquake prediction, the deformed frog mystery, and global warming. Each learning activity begins by engaging students in questions that assist teachers in ascertaining what knowledge they bring related to the assigned topic. After students reflect upon their current understandings, they are immediately connected to learning about and responding to contemporary scientific controversies. Throughout the activity, students are continually evaluating information from predetermined websites and recording that information in an online journal. In the end, students review the information they saved, color-code themes from the data, and construct an argument based on these themes in order to design debates to support their position. All of the interactions take place through and on the Internet.

MULTIMODALITY AND THE CHALLENGE OF ENGAGING DIVERSE WAYS OF KNOWING

Students commonly misunderstand the discipline of science as one that is a collection of facts or truths. This misunderstanding is often reinforced by textbook designs, assessment activities, and classroom practices that emphasize unquestioned memorization of scientific concepts. While arguably important to the science curriculum, these concepts need to be understood in both the context of learning and the sociocultural-historic practice in which they were constructed. The practice of science is characterized by competing hypotheses, emergent data, and continually revised scientific explanations. Within these debates, the assumption that knowledge is rational, neutral, and universal has been privileged over the view of knowledge as mediated by experience and context. These privileged ways of knowing have contributed to the marginalization of women and communities of color within the practice of science. Roth (Chapter 8) argues that an important goal for scientific literacy is the capability to stand up for and fight for one's rights and interests in contested issues. Central to this goal is finding ways for students to engage diverse ways of knowing within this practice.

Scientific explorations using different data forms engage diverse ways of knowing. Advances in interactive, computer-based animations and visualizations create *multimodal* opportunities for students and teachers to see, understand, and explain complex science concepts. These emerging technologies create opportunities for students to learn through *new* modalities as well as construct understandings *across* modalities. That is, image, text, numbers, and even touch-based data can be considered together within a science exploration where students read texts (including images), manipulate 3-D visualizations, and identify the relationship between qualitative and quantitative data. One specific example of a way technology has increased the opportunity for students to learn through new modalities is *haptics*. Haptics, learning through touch, has been applied within nanotechnology education through the use of atomic force microscopy. This new technology provides opportunities for physical, touch-based manipulation at the nanoscale within science explorations. For example, students can explore the mystery of a sick puppy by experimenting with the capsid (a protein shell that surrounds a virus particle) to determine the shape of the virus, test for DNA or RNA, and determine the size of the virus to make a diagnosis.

In other tools, students are encouraged to explore and construct understandings across multiple modalities. For example, in *WorldWatcher* students are supported in their investigations of global warming through interactive map software or geographical information system (GIS). In this software, students have opportunities to match their understanding of climate to the color patterns they see on the map. These color- or image-based patterns are representations of essential numeric data that are also available for student exploration.

Stella and Model-It are two platforms that use modeling and simulation to illustrate varying arrangements of data. These platforms help students to construct and test models, which provide opportunities for students to view data from multiple data forms while they develop understandings of natural systems.

Students can easily build, test, and evaluate qualitative models without having a background in the complex mathematical computations driving these models. Students can also customize and link representations that simulate their own environment. The customized model represents the students' theory about how variations in particular scientific phenomena introduced in class interact in the real world. Following the construction of a theory, students create interacting connections between the variables in the environments, which model the causal relationships before their very eyes. Students have the ability to easily change the values of a given variable in the model and immediately see the effects of that change through meter and graph data visualization tools provided within Model-It.

CONTEXTUALIZATION AND THE CHALLENGE OF BRIDGING SCHOOL/OUT-OF-SCHOOL PERSPECTIVES

The practice of science is enacted across a spectrum of individuals, communities, and institutions in which knowledge is produced and acted upon. The forms of knowledge contributed through these efforts vary, reflecting different dispositions toward and understandings of scientific practice. These differences contribute to debates around possible ecological, economic, political, and social impacts (Roth, Chapter 8). Scientific literacy ultimately involves the ability to participate in debates by bridging perspectives from a range of school and out-of-school contexts. Two implications are argued here. First, students must have opportunities to be active contributors to scientific discourse by actively collecting data and constructing their own explanations. Second, students need to be able to contextualize their own scientific practice in relation to the practices of others. Becoming fluent with the use of emerging technologies is one supportive means of facilitating these practices.

There is a range of software applications that assist in the organization and analysis of collected data so that students may contribute to scientific discussions. Two examples are WISE and Kids as Global Scientists. The WISE (described above) interface has embedded within it a function that allows students to peer review other students' notes and debates both within a given class and across the WISE community. Similarly, Kids as Global Scientists has the purpose of allowing many participants to view the same data. Kids as Global Scientists is an inquiry-based science program that engages students, teachers, and parents in collaborative debates. In these programs, participants use the same weather data from the Internet, along with archival weather data. The activity centers on investigating weather and climate concepts in their city and supporting collaboration between students and science experts around real-time and archived weather and species data sets. More open-ended options include iEARN, Global SchoolNet, and ePALS, which provide supports for identifying and collaborating with groups of students around the world to conduct collaborative investigations. These programs offer a space for collaborative online interaction and do not provide predesigned inquiry activities.

While the above examples elucidate ways in which the Internet supports increased opportunity for contextualizing science inquiries within larger, global debates, emerging technologies also create opportunities for students to contextualize science inquiries within local phenomena and data. Handheld technologies, for example, facilitate dynamic, real-time data collection and sharing across school and community contexts. Imagine a group of sixth graders studying ecosystems. They are walking to the creek two blocks from their school. Using probes connected to their handhelds, students test the pH balance of the water at different points along the creek and are able to instantly plot these data into a spreadsheet for later analysis. Students are then encouraged to practice their skills of observation as they use a camera attachment with their handheld to take pictures along the banks of the creek. Students use the word-processing function of their handheld to write about what they see. Back in the classroom, students beam their collected data to one another. At home, students do a mindmap of questions about the creek watershed and local ecosystem that they want to explore further. In school the next day, students beam this homework into the teacher, who is able to draw from these ideas to identify connections between students' questions and the targeted science concepts of the ecosystem unit. Students then design procedures for collecting further data to inform their questions. The mobility of this technology facilitates collection and examination of data across different contexts. Students are no longer bound by physical constraints of wires, and thus are able to contextualize data both inside and outside the classroom walls. Students can then publish their findings on the Internet. While there are many software applications available specifically for web page design, Microsoft Word and Claris Works are examples of more ubiquitous software that can easily convert documents into web page format. In this way, students can easily connect their classroom-based science explorations to a wider audience.

TAPPING YOUTH AS A RESOURCE TO BRIDGE THE TECHNOLOGY LITERACY DIVIDE

While calls for technology fluency continue within science (and across the curriculum), evidence suggests that computer technologies in schools continue to be underutilized. This is explained in part by reports from teachers that they feel underprepared and undersupported in helping students to develop these new literacies. Despite continued reports of a "digital divide" among other groups, 90 percent of children between the ages of five and seventeen now use the computer in home, school, or community contexts (U.S. Department of Commerce 2002). Youth, particularly those living in low-income neighborhoods, have turned to community-based organizations as an alternate site for technology access and learning. Not only are teens the "early adopters" of these technologies, surpassing adults in the use and integration of these tools in their daily lives, but they also represent an increasingly targeted consumer market for which a host of new products is being exclusively developed. In addition to the burgeoning online commercial culture centered around youth popular culture and consumerism, there is a growing web culture of youth expression and activism in political, cultural, and civic life.

Typically, youth have little input into discussions around why and how technology should be used in school. Teachers and administrators make these decisions, often having limited information on whether youth have access to technology outside school and what they do with technology when they do have access. The risk, we argue, is that youth come to school with technology experience and expectations that may be overlooked or even undermined within the classroom. The result is a growing disconnect between teachers and their more tech-savvy students (Levin and Arafeh 2002). Left unattended, this disconnect is likely to impede opportunities for learning and contribute to a pattern of disengagement in which some learners fail to see schooling as an avenue for life progress. Ultimately, we need new strategies for tapping youths' out-of-school experience and practice as an available resource for teaching and learning in the classroom.

REFERENCES

International Technology Education Association. 2004. *The Second Installment of the ITEA/Gallup Poll and What It Reveals as to How Americans Think about Technology*. www.iteawww.org/TAA/PDFs/GallupPoll2004.pdf.

Levin, Doug, and Sousan Arafeh. 2002. *The Digital Disconnect: The Widening Gap between Internet-Savvy Students and Their Schools*. Pew Internet & American Life Report. Washington, DC: Pew Charitable Trusts. www.pewinternet.org/reports/toc.asp?Report=67.

National Academy of Engineering. 2002. *Technically Speaking: Why All Americans Need to Know More about Technology*. www.nap.edu/books/0309082625/html.

U.S. Department of Education. 2004. *Toward a New Golden Age in American Education: How the Internet, the Law, and Today's Students Are Revolutionizing Expectations*. National Education Technology Plan. www.nationaledtechplan.org/docs_and_pdf/National_Education_Technology_Plan_2004.pdf.

ADDITIONAL RESOURCES

Suggested Readings

Baumgartner, Eric, and Phillip Bell. 2002. *What Will We Do with Design Principles? Design Principles and Principled Design Practice*. http://scale.soe.berkeley.edu/papers/BaumBell2002.pdf.

DiGiano, Chris, Louise Yarnall, Charlie Patton, Jeremy Roschelle, Deborah Tatar, and Matt Manley. 2002. *Collaboration Design Patterns: Conceptual Tools for Planning for the Wireless Classroom*. Center for Technology in Learning, SRI International. Proceedings of the IEEE International Workshop on Wireless and Mobile Technologies in Education (WMTE '02). http://ctl.sri.com/publications/downloads/DiGianoWMTE.pdf.

Jackson, Shari L., Steven Stratford, Joseph S. Krajcik, and Elliot Soloway. 1995. *Learner-Centered Software Design to Support Students Building Models.* http://hi-ce.org/papers/1995/learner_centered/index.html.

Luchini, Katy, Chris Quintana, and Elliot Soloway. 2003. Evaluating the Impact of Small Screens on the Use of Scaffolded Handheld E-Learning Tools. *Journal of Computer Assisted Learning* 19 (3): 260–72. http://hi-ce.org/PDFs/LuchiniAERA04.pdf.

Roschelle, Jeremy. 2003. Unlocking the Learning Value of Wireless Mobile Devices. *Journal of Computer Assisted Learning* 19 (3): 260–72. http://ctl.sri.com/publications/downloads/UnlockingWILDs.pdf.

U.S. Department of Commerce. 2002. *A Nation Online.* Washington, DC.

Websites

U.S. Department of Education, Office of Technology in Education: www.ed.gov/about/offices/list/os/technology/index.html

Technology supports listed within the "Scientific Inquiry and the Challenge of Navigating Available Data" section, above

Ask Jeeves for Kids: www.ajkids.com
KidsClick!: http://sunsite.berkeley.edu/KidsClick!/
ThinkQuest Library: www.thinkquest.org/library/
Web-based Inquiry Science Environment (WISE): wise.berkeley.edu
WebQuests: http://webquest.sdsu.edu
Yahooligans: http://yahooligans.yahoo.com

Technology supports listed within the "Multimodality and the Challenge of Engaging Diverse Ways of Knowing" section, above

Model-It: http://goknow.com/Products/Model-It/
Nanoscale Science Education: Learning through Touch: http://ced.ncsu.edu/nanoscale/haptics.htm
STELLA: www.iseesystems.com/(200xvdfinx11i245qpohtsu2)/softwares/Education/StellaSoftware.aspx
WorldWatcher and the GEODE Initiative: www.worldwatcher.northwestern.edu

Technology supports listed within the "Contextualization and the Challenge of Bridging School/Out-of-School Perspectives" section, above

ePALS: www.epals.com
Global SchoolNet: www.globalschoolnet.org/index.html
HICE Handhelds HomePage (Center for Highly Interactive Computing in Education): http://handheld.hice-dev.org
iEARN: www.iearn.org
MIT PDA Participatory Simulations Site: http://education.mit.edu/pda/index.htm
One Sky, Many Voices (Kids as Global Scientists, BioKids, Hurricanes): www.onesky.umich.edu

PART 3

HOME AND
SCHOOL RELATIONSHIPS

15

Elementary Students and Parents Learning Science through Shared Inquiry

Judith A. McGonigal

According to the *National Science Education Standards,* "inquiry into authentic questions generated from student experiences is the central strategy for teaching science" (National Research Council 1996, 31). The fundamental abilities necessary to conduct science inquiry in kindergarten through fourth grade, as identified by the National Research Council (2000, 19), are as follows:

- Ask a question about objects, organisms, and events in the environment.
- Plan and conduct a simple investigation.
- Employ simple equipment and tools to gather data and extend the senses.
- Use data to construct a reasonable explanation.
- Communicate investigations and explanations.

Teaching science through inquiry has proven very difficult for many elementary grade teachers. To create authentic and effective science inquiry, teachers can enlist the support of students' parents. Below, I discuss several examples of how parents collaborated with my first-grade students to conduct shared science inquiry. Each of these examples features long-term investigations that introduced students to science inquiry. With each project, we expanded the role of parents and home resources in supporting the learning of science through inquiry.

INTRODUCING SHARED INQUIRY IN THE CLASSROOM

Our first shared inquiry involved collaboration between students and parent volunteers. During the first month of the school year, students planted raw peanuts with and without shells in a classroom grow lab. With the help of a parent volunteer, students observed the peanut plants germinating and growing, and documented their observations in a science notebook. Later, the first graders designed experiments to determine which variables influence plant growth. With the help of parent volunteers, they gave some of the plants chemical additives and varied the amount of sunlight that plants were exposed to.

As scientists work with materials, new questions emerge. This was true in our classroom as well. When flowers appeared on peanut plants, the first graders wondered if the flowers would produce new peanuts. One of the parent volunteers called a local farmer for information about how to pollinate the flowers. The farmer invited the first graders to his farm. With six parent chaperones, the first graders learned about the reproduction of legumes and harvested raw peanuts from the farmer's fields.

This first shared inquiry provided the initial framework for a model of science inquiry with parental support.

EXPANDING THE ROLE OF PARENTS

For our next class project, the role of parents in shared science inquiry was expanded. During the second month of school, I set up several terrariums in the classroom. I ordered bess beetles, Madagascar hissing roaches, and a scorpion from a biological supply company. At a pet store I purchased mealworms. The goal of this project was for students to discover how small creatures live, grow, and reproduce.

This classroom project quickly became a management challenge. Students needed time and supervision to feed the insects, clean the cages, and record their observations. To resolve this issue, I recruited five parents to volunteer one day a week. Each day, one parent would stay after school with a group of students and help them care for the insects. During this time, students also recorded in words and pictures what they had noticed and what problems they had solved. Together, the parent volunteers and the children constructed knowledge about the insects and their environments as they handled the materials and gathered evidence about the physical world.

We also planned a family field trip to extend this shared inquiry. In response to an invitation I sent to the first-grade families, eighteen students and their parents met me at the Insectarium, a museum in Philadelphia, Pennsylvania. Together, parents and children learned how to mount butterflies, watched a video about insects' life cycles, and listened to a lecture about how to provide for the needs of insects housed in terrariums.

This second inquiry extended the original model of shared inquiry with parental support to include greater parental involvement in the classroom and increased parental participation in a shared-inquiry field trip.

WATCHING A MODEL FOR SELF-SELECTED
SHARED INQUIRY EMERGE

Authentic inquiry is fueled by the learner's passion to answer self-generated questions. I decided to move from teacher-directed science inquiries to student-selected inquiries. During the third month of school, students were invited to bring into school items that they thought might be interesting for classmates to observe. Students brought a hermit crab, a large maple leaf that had turned crimson red, a horseshoe crab shell, and an old baseball that had been opened up to reveal the inside layers.

The internal structure of the old baseball piqued even the teacher's curiosity. I wondered what was inside other balls. I brought to school an old golf ball, and asked the student who had opened the baseball to take the golf ball home and find out what was inside. To both our amazement, he returned the next day with a collection of rubber bands that he had found inside the golf ball. He explained that when he and his dad cut the golf ball open, it seemed to "spring to life" as the rubber bands unwound.

Following the public sharing of this discovery, students began bringing additional balls to the first-grade "ball expert" to take home and open. Over time, the ball inquiry expanded to include several tests before dissection: a ball bouncing test, which was carried out from the top of his clothes dryer; a ball rolling test on the cellar floor; and a careful observation and recording about the texture of the ball's surface.

After working at home for a month with both parents, the first-grade ball expert presented his findings to his peers. He attached his evidence, eighteen dissected balls, to a whiteboard and brought it to school. His classmates listened, questioned, and challenged the ball expert as he explained how the texture and internal structure of the ball influenced how the ball rolled, bounced, and flew through the air. I listened and watched as a model emerged for having first graders conduct self-selected science inquiries in the home.

That very day I asked each first grader to select a topic to investigate, as they had seen the ball expert model for them. Without hesitation, the twenty 6-year-old students selected their own topics. Many of the choices were inspired by materials in the classroom, and included flowers, fish, frogs, hair, clocks, symmetry, crayons, leaves, and Madagascar hissing roaches.

INFORMING PARENTS ABOUT SHARED
SCIENCE INQUIRY IN THE HOME

At my request, the ball expert's mother composed a tip sheet for families to use as a reference guide for home science inquiries.

1. Select object, idea, or concept of great personal interest.

2. Gather as many different kinds of your object of study.

3. If applicable, "dissect" your object.

4. Make comparisons to discover similarities and/or dissimilarities.

5. Note different properties of your object.

6. Measure its capabilities/aspects/characteristics.

7. Try out experiments with your object.

8. Make predictions along the way, and try to prove or disprove them.

9. Let the *inquiry* evolve. Don't try to do it all at once.

Created by Marvin's mother

FIGURE 15.1   Ideas for inquiry investigations.

Along with this parent-generated list of directions, I sent home with each student copies of two chapters from *Primary Science...Taking the Plunge* (Harlen 1985). These chapters provided detailed information on the essence of inquiry—how to help children observe, raise questions, and answer them. I also sent home a personal invitation for each family to support their child's self-selected science inquiry.

Dear Families of First Graders,

Our next presentation is an individual science inquiry. Each student will be asked to share with the class what s/he has learned while doing his/her science inquiry. The presentation should report three things:

What question did the student try to answer?

What things did the student do to try to find the answer?

What did the student learn?

After telling us the answers to these questions, classmates will be invited to ask the student presenter questions in order to probe for more facts.

Students should have some type of display of materials to help show how they did their inquiry. We are not looking for print material as a source of information. We are looking for our young scientists to learn from handling materials. Presentations should be done by March.

Students will probably need adult support, but please do not take control of the investigation. Let the child and the materials lead the inquiry.

Parents can:

Help obtain materials,

Assist in using any dangerous tools,

Provide an audience for discoveries,

Help schedule time to work on the inquiry,

Help the student find a way to record observations,

Call me with any questions or concerns.

We are as much concerned about the process of doing an individual inquiry as we are with obtaining sophisticated scientific understandings. This will only be a beginning of perhaps a lifelong inquiry for your child. There will be, as in all scientific investigations, much more to learn and discover. The important fact is that each young scientist learns that s/he can discover information by handling and observing something.

Thanks again for all your support. If I can assist you in any way with the science inquiry, please call me. Please do not become stressful about the science inquiry. It is just an opportunity for your child (and perhaps your entire family) to experience the joy of learning, the joy of doing science.

Hope to hear from you with your questions,

—Judith A. McGonigal

CREATING A FEEDBACK LOOP FOR COMMUNICATING PROGRESS

Parents did not immediately respond to the invitation to engage in a science investigation with their child. No one contacted me about questions, concerns, or accomplishments. I decided that I needed to create a feedback loop for the families and students to report how they were conducting science inquiry at home.

Response Sheet: Week 1

I developed a science notebook response sheet that each family, as part of the weekly assigned homework, needed to complete. Student scientists were encouraged to dictate their answers to an adult reporter.

The feedback that I received revealed that many students had not yet started their self-selected inquiry because they were unable to get the materials needed.

Scientist: _____

Reporter: _____

What material or topic are you observing and investigating?

What have you done so far as part of your investigation?

What have you learned?

What problems are you facing?

When can your teacher visit to learn about your investigation?

FIGURE 15.2 My first-grade science inquiry.

Sharing Resources: Week 2

To resolve the issue of materials, I generated a letter listing each student's investigation and requesting that families support each other in acquiring the needed resources.

Feedback on this response sheet informed me that classmates had begun supporting each other's investigations. Feathers had been collected on a nature walk and given to Carolyn, hair samples from family members had been provided to Raymond, and a log from one student's garage had been delivered to Thomas's house.

As Fruchter, Galetta, and White (1992) have previously documented, parents are key resources for improving not only their own child's education but also the schooling of other children. The invitation to support other families in their science investigation made this readily visible to the students, their parents, and me.

Revising the Response Sheet: Week 3

I wanted to communicate more clearly to parents and students that a science inquiry often starts with questions, and includes identifying and solving problems. For the third week of the home inquiries, I created a new science notebook response sheet that focused on questions, problems, and solutions.

The responses to this science inquiry notebook worksheet were very informative. I learned about family trips, including a visit to the Crayola Factory to observe crayons being made, a visit to an ecological center to view the composting of leaves, and a visit to an architect's office to learn about forces that affect structures.

Students and parents reported problems that they had solved, including the use of gloves to protect hands when investigating cacti, the development of a labeling system for crystals, and the relocation of potted flower seeds on a dishwasher to provide additional heat and moisture.

Each family followed its own course to making sense of science inquiry. As described by Wells (1994), the use of the notebook response sheets allowed me to respond with family-specific guidance that matched each family's shared zone of proximal development.

PROVIDING TOOLS TO SUPPORT SHARED INQUIRY IN THE HOME

To support the families and to continue to move the discourse for science into students' homes, I delivered tools to help facilitate the investigations. Several students, who discovered that their toy microscopes were ineffective, were invited to borrow the high-quality microscope that I had purchased at a flea market. I lent a biological catalogue to a family who wanted to obtain hornworm caterpillars for their son to observe. The frog expert borrowed a

Student's Name_____

Parent's Signature_____

Please report back to me on how you have supported other learners.

Continue working on your individual science investigation.

Complete the enclosed science notebook page by yourself or dictate your answers to your parent to record.

If you have any material or information that would help any of the other student scientists in our classroom, please share it with them. We want to become a community of learners that support each other's investigation.

Raymond is studying hair. He needs samples of long hair to test for strength.

Ruth is studying crayons. Does anyone know about visiting a crayon factory?

Thomas has changed his topic from insects. He is now studying tree trunks.

Philip is studying clocks. He needs analog and digital clocks to take apart.

Rebecca is studying pencils.

Walter is studying mealworms.

Robert is studying symmetry. Can anyone build or bring him symmetrical items?

Dorothy is studying cactus. Where could Dorothy go to see an outstanding collection?

Neal is studying caterpillars. He should receive his live specimens in the mail this week!

Mary is studying our hissing roaches.

Marguerite is studying cat behaviors. Do you have a cat that she might come and observe?

Danielle is studying flowers.

Carolyn is studying feathers. Do you have any uses for feathers in your house that she could observe?

Paul is studying rocks. Gary's Gem Shop on Route 70 is a great place to visit to learn about rocks.

Denise is studying structures.

Stanley is studying leaves. I gave him some of the leaves you pressed earlier this school year.

Allen is studying fish. Does anyone have a tropical fish tank in operation? Perhaps he could observe it and interview you.

Daniel is studying crystals. I'm reading a biography about Marie Curie. Did you know that the study of how crystals are made was her husband's early investigation?

I can't wait to learn what everyone is learning.

FIGURE 15.3 Your work with others.

MY SCIENCE NOTEBOOK

Scientist:_____

Reporter:_____

What did I do for my investigation this week?

What new questions have come to my mind as I work on my investigation?

What problems am I having?

What investigation problems have I been able to resolve this week?

What did I learn this week while working on my investigation?

---

terrarium when his tadpoles needed a larger habitat. A digital balance scale was lent to the crayon expert, who was trying to discover if the mass of a crayon changed after it had been melted and rehardened. Sharing my science resources with families continued to blur the boundaries between home and school, and provided additional opportunities for parents and child to extend science discourse in the home.

PARENTS ENGAGING IN SCIENCE DISCOURSE

Once families began engaging in home science inquiry, science discourse became an important part of everyday life. At scout meetings and movie nights at school, I heard dads talking to each other about their child's science inquiry:

> Did you know at our house fungus is among us! Thomas has become more interested in the fungus on those logs than the rings.

> Can you believe my wife beeped me on my pager at work when the first beetle emerged from its pupa?

Likewise, I overheard mothers, waiting outside the school for their children, talking science:

> Did you know that the melting point of Styrofoam is lower than the melting point of crayons?

> My son has developed better recipes for rock candy than the one in the crystal kit I purchased at the store.

Together, adults and children were learning by handling the materials and interacting with each other in the home. Not only the first graders, but also the parents, were beginning to connect the discourse of science with their primary discourse.

COMMUNICATING SCIENCE NEWS TO AN AUDIENCE

After the families had been carrying out their home science inquiries for four weeks, I scheduled a home visit for each student. I initially scheduled home visits as a means to hold families accountable to support their child. I wanted to communicate to families how much I valued the self-selected inquiries, and I believed that I was going to identify many students who were having difficulty carrying out their science investigation.

Instead, I discovered that the inquiry assignment had created a time and place for each family to explore science together. My home visits became an occasion to celebrate the families' discoveries, and to challenge each family to extend its investigation.

The visits typically began with the student escorting me to the site of science inquiry in the home. Crystals were grown on top of the radiator in Daniel's dining room. Straw structures filled every corner of Denise's family room. Cactus plants flourished on Dorothy's heated back porch. The pupae of hornworms metamorphosed inside Neal's bedroom closet, and Raymond's hair strength tests were carried out on his kitchen table. Frogs swam in Steven's bathtub, and Stanley's leaves were collected and displayed on the dining room table.

With the evidence of their research in front of them, parent(s) and child told me what they had done and what they had learned. I had a clipboard and camera to document the visit. As family members were sharing and explaining, I carefully took notes and photographs. The notes were later used to guide the first-grade scientists during their classroom presentations if they became distracted by the challenge of speaking to a large audience for a significant amount of time.

At each home visit, I often paraphrased what the child said, to clarify for the child, the parent(s), and myself how the student was making sense of the phenomenon under investigation. When appropriate, I introduced new science vocabulary that matched the child's reported observations.

The home visit ended with teacher-directed suggestions for at least one additional activity for the family to try before the classroom oral presentation. The symmetry expert was shown how to use a mirror to test for symmetry. The pencil expert was encouraged to visit an art supply store to see how pencils were labeled with numbers. The leaf expert was challenged to replicate his results of forcing buds to open indoors. The rock expert was given a geode to open.

SHARING KNOWLEDGE WITH PEERS

Beginning in February, each student selected a day to share his or her science inquiry with the class. The child, with the help of a family member, set up a presentation, which usually included a whiteboard, artifacts, and tools used to do the research. Each morning, first graders entered the classroom and eagerly approached the presentation display table to examine the new artifacts. There was a shared expectation that each day we would construct new science understandings by interacting with classroom peers and new materials.

At 11:00 A.M., the featured young scientist would share his or her science news and respond to questions from classmates and the teacher. After the presentation, which lasted at least twenty minutes, each member of the audience was required to complete a science notebook report that answered three questions:

1. What did the scientist do to study the material?
2. What did you learn from the scientist?
3. What questions do you now have that could be investigated about this topic?

The written responses, which were used to encourage active listening, were individually read to the science presenter and then compiled into a booklet, which was given to the presenter to take home and share with family members.

EXPANDING THE AUDIENCE

Parents, pleasantly surprised and proud when they read the feedback sheets, began voicing their regrets that they had not been present to see their child's presentation. After a lengthy class discussion, the first graders decided that each classmate should be given an opportunity to schedule a second presentation for their parents and another invited class. Many first graders selected this option and were given a second chance to showcase their science expertise.

As more and more presentations were given, students started to make connections between the inquiries. The life cycle and the horn used for defense of the tobacco hornworm were compared to the life cycle and the horn-like appendages on the abdomen of the mealworm. Numbers used to identify the softness of lead pencils were viewed as similar to the numbers on the rock expert's Moh's scale for hardness. Robert began to spot symmetry in the materials displayed for each presentation.

PARTICIPATING IN A SCIENCE INQUIRY FORUM

On the last school day in March, we held a Science Inquiry Forum in our classroom to celebrate our science learning. Every child displayed his or her presentation board and artifacts, along with a student-authored book about the inquiry. In the evening, students gave their families tours around the Science Inquiry Forum and explained the science inquiry that each display showcased. Parents were able to see the results of each family's science inquiry, and were able to appreciate the tremendous resource for learning that had been tapped when families enriched the learning opportunities available in the classroom.

The breadth and depth of science content shared that evening provided additional evidence of how students and parents could successfully function at home and at school as colearners of science through inquiry. When parents and teacher worked together, they were able to discover a way for first graders to meet the *National Science Education Standards* of learning science through inquiry (National Research Council 1996).

THE LASTING EFFECTS OF SELF-DIRECTED SHARED SCIENCE INQUIRY

A year later, I received the following e-mail from a student's mother:

As a family, we still look for symmetry in our everyday lives and often use the "mirror test" to confirm our findings. Robert still makes comments about different things he learned from other children's science projects, such as crystals and hair. He connects it to what he is doing and seeing. As a family, we have become so much more aware of science, and we enjoy it!

REFERENCES

Fruchter, Norm, Anne Galetta, and J. Lynne White. 1992. *New Directions in Parent Involvement.* Washington, DC: Academy for Educational Development.

Harlen, Wynne, ed. 1985. *Primary Science . . . Taking the Plunge.* Oxford: Heinemann Educational.

National Research Council. 1996. *National Science Education Standards.* Washington, DC: National Academy Press.

———. 2000. *Inquiry and the National Science Education Standards.* Washington, DC: National Academy Press.

Wells, Gordon. 1994. Learning and Teaching "Scientific Concepts": Vygotsky's Ideas Revisited. URL: education.ucsc.edu/faculty/gwells/Files/Papers_Folder/Sci entificConcepts.pdf.

ADDITIONAL RESOURCES

Lawrence-Lightfoot, Sara. 2004. *The Essential Conversation: What Parents and Teachers Can Learn from Each Other.* New York: Random House.

Saul, Wendy, and Jeanne Reardon, eds. 1996. *Beyond the Science Kit: Inquiry in Action.* Portsmouth, NH: Heinemann.

Tobin, Kenneth. 2001. Learning to Teach Science as Inquiry. In Wolff-Michael Roth, Kenneth Tobin, and Stephen Ritchie, *Re/constructing Elementary Science,* 276–307. New York: Peter Lang.

16

Rethinking Homework

Jennifer Beers

For many teachers, student failure to complete homework assignments is a pervasive problem that can create obstacles for learning and achievement. One solution to this problem is to not assign homework every day or, in the most extreme case, completely do away with this type of assignment. However, in our current state of high-stakes schooling, standards-driven curriculum, and teacher accountability, I do not believe that any educator can ignore that homework, if framed in the proper context, can be a very useful tool. It is extremely useful for practicing the skills and concepts required to master course content. Additionally, in science, it can also be a way to link the abstract concepts and ideas learned in class to students' personal lives, families, and communities.

Given this dilemma, there are several questions that can be raised. What can educators do to improve the likelihood that students will complete their homework? How can homework be structured so that students will be able to complete it? What should be taken into consideration when planning the forms of assessment that students are required to complete? In this essay, I explore the ways in which I have attempted to improve student submission of homework assignments. First, I will describe how inquiry into my own practice has framed the ways in which I construct many "out-of-class" assignments. Second, I suggest that inquiry into the students' lifeworlds and the structures that impact their ability to complete assignments is crucial for understanding the "homework dilemma." Throughout this discussion, I highlight some of the more "macro" structures (school or department policies, state standards, etc.) that impact a teacher's ability to best serve her students with regard to completing homework.

WHY AM I GIVING THIS ASSIGNMENT?

Teachers face many pressures with regard to their curriculum and pedagogical approaches in the classroom. These pressures can come from within the school administration or from outside influences like school districts or state and national standards. Moreover, teachers must negotiate their own experiences from teaching and learning in other schools or contexts. All of these outside factors and "macro" structures can impact the way they approach assigning homework as well as how they view its importance.

In working for four years in an urban charter school, I have grappled with these issues and tried to work out an approach to homework in a way that would serve both the needs of my students as well as the requirements set up by the school. This school strives to create a culture of high expectations so that our students can be competitive with those from other schools. In this regard, there has always been a clear message from administrators that students must be given two to three hours of homework a day in their core subjects to be competitive. In an effort to support the school administration, I struggled to find homework assignments for my students to complete every evening. Oftentimes, I found myself creating assignments simply for the sake of giving my students something to do when they left the building. Thus, some of these assignments lacked the purpose, vision, and connection to the science being taught in class. Students often failed to complete work because I did not make its purpose explicit. Furthermore, I would often attribute their failure to submit homework to their abilities or understanding of course content, rather than to their motivation and desire to complete the assigned task.

At this school, success and failure rates at midterm grade reporting are often the catalyst for a review of classroom practices and schoolwide structures. Several professional development sessions are dedicated to the topic of homework and its completion. These sessions have been extremely helpful with regard to learning from more experienced teachers and sharing successful strategies and ideas. Throughout these sessions, I reflected on the philosophical struggle between what I know is in the best interests of my students and what I felt was necessary to be in good standing with my colleagues and supervisors. It has become clear that if there are circumstances in which homework is given merely for the purpose of giving work to students to complete outside of school, then one must rethink her philosophy on such assignments.

In addition to schoolwide pressures, the scope and breadth of the content required by the state standards are especially daunting (Pennsylvania Department of Education 2004). While there is pressure to help students to be successful on standardized tests, it is important to note that in a yearlong course, it is almost impossible to teach everything students need to know for these tests. An immense pressure to forge through the curriculum creates contradictions between high expectations for student achievement and realistically helping students understand the most important pieces of course content. Personally reviewing the quality and quantity of homework assignments with both my students and other teachers has helped me to deal with my naïve conceptions of what is absolutely necessary for students to know in my courses. While most of us model our own teaching practices after previous experiences with science, the scope and breadth of state science standards do not require us to cover every minute detail; rather, they necessitate that we help students understand overarching themes. For example, a teacher might ask, "Is it crucial for my students to know the names of the mitotic stages or is it more important that they know the consequences and outcomes of the process?" By asking these types of questions, I have been able to pare down my curriculum so that students are able to get a well-rounded experience with science that includes the reading and writing skills required to be successful on standardized tests. Moreover, this has relieved some of the pressure to assign meaningless homework and has created opportunities for students to demonstrate their understanding of the big themes and practice the more generalized skills required for state tests.

Solutions to the "homework horror" have been written about extensively (National Education Association 2005); however, all successful educators have their own structures in place to improve the quality and the quantity of homework assignments submitted by students. At the end of each semester, I ask my students to complete a homework survey. In addition, inquiry and reflection on the origin of my personal homework philosophy as well as my practices on assigning homework have helped me to generate the following conclusions/ideas:

1. Students are more likely to complete homework assignments if they feel that they will be given a grade or credit for their work.

2. Students are more likely to complete homework assignments if they understand the purpose of the assignment and can clearly make a connection between the task assigned and their daily learning.

3. Teacher educators and mentor teachers should help new teachers work through any naïve conceptions of what students absolutely needed to know and what constitutes "homework."

4. Schools should create opportunities for teachers to formally share their own philosophies and ideas about homework, including examples of successful strategies and assignments.

HOMEWORK SHOULD BE DONE AT HOME, OR SHOULD IT?

In addition to the pressures outlined above, the dilemma over homework can be exacerbated by a teacher's failure to take into consideration the impact of many cultural, social, and economic issues that affect students' ability to complete work at home. As a teacher, I believe that, in most cases, overlooking these factors is almost unintentional as many of us come from vastly different backgrounds as compared with our students. The way we were brought up, the support we received at home, and our obligations after school could have been very different from those of the youth in our classrooms. Hence, the notion that "homework should be done at home" needs to be reexamined and rethought.

Previously, I have struggled with the idea that homework should be completed outside of school without truly understanding my students' lifeworlds. I used my own experiences at home and in school as a starting point for my practice without realizing that my students face different pressures and obligations. It was not until I started to conduct regular cogenerative dialogues (see Chapter 31) with students that I became aware of many of the constraints that face my students. In candid conversation, they cited afterschool jobs, taking care of younger siblings, housework, the extended hours of our school day, and the length of their commute home as reasons for their inability to sometimes complete homework assignments. Some of my students also stated that they feel safer at school and, therefore, like to stay after to socialize with friends and complete work. Furthermore, they often described cramped and noisy living conditions, which further prevented them from being able to concentrate.

As a result of these dialogues, one of our cogenerated actions required me to rethink when and where students complete their homework. Thus, I began inviting students to stay for after-school office hours. I thought that if I want my students to complete assignments, then I should make a space for them to complete their work and be successful in class. I also made an effort to come to a better understanding of my students' lives so that they might be able to negotiate due dates for work. Also, I began to rethink the assignments themselves. For example, "new" assignments ask students to interview someone about asthma, write a rap/poem about molecular movement, or do a simple experiment at home.

Rethinking homework has been a catalyst for me to take a more holistic approach to this teaching tool. In this regard, I have come up with the following suggestions that involve parents, teachers, students, and schools in improving student achievement with homework:

1. Teachers could set up cogenerative dialogues so that they see the impact of outside pressures/obligations on student performance.

2. Schools could set up a homework club staffed by parent volunteers, student interns, or older students.

3. Teachers and parents could construct ways in which they communicate about homework and its completion (e.g., the planner system or e-mail).

4. Teachers could tailor their assignments so that there may be more adult involvement.

5. Students could find ways to communicate with their teachers about the problems they face with completing homework.

SAMPLE HOMEWORK ASSIGNMENT

In this section of the course, you will be learning about what happens inside an individual who has asthma. Asthma is a condition that affects a high number of people who live in cities around the United States. Some of you may also be living with this respiratory condition. If not, then many of you probably know someone in your family or your community who deals with asthma everyday.

In this assignment, you need to interview someone you know who has asthma and write a three-paragraph (3–5 sentences each) summary about what you learned from your interview.

Alternative: If you have asthma, you have the option of writing a three-paragraph reflection on what it is like to live with the condition.

Brainstorm Your Questions: What do you want to know? What do you want to learn as a result of your interview? In the space provided, brainstorm the questions you want to ask during your interview.

Conduct the Interview: Use your questions to help guide you through the interview. Make sure you take some notes on a separate sheet of paper so that you can remember what the person has said. Or, if you have a tape recorder, you could record the conversation.

Write Up the Interview: On a separate sheet of paper, write up what you learned in the interview. You should also pose other questions and add a personal reflection on what it must be like to live with asthma.

FIGURE 16.1   Asthma interview.

FINAL THOUGHTS

The "battle" over completing homework is certainly not new or restricted to the particular cohort that I teach. Also, its function in the education of young people is timeless. It can be a mechanism for practicing skills, reviewing material, and reflecting on learning. It can help teachers understand the holes in student understanding and can be a valuable assessment tool. However, when homework is used simply to give students work to do at home or because teachers feel pressure (real or perceived) to assign it, then the utility of the task can be lost.

In this regard, I urge teachers to reflect on the origin of their practices relating to homework, as this might open new avenues for constructing and assigning such assignments. I challenge students to open up to their teachers if they have particular pressures or problems that prevent homework completion. Also, I ask parents to continue open lines of communication with

Use the following words in a creative way to show the differences between the different types of molecular movement that we covered in class. You will be required to submit a paper copy of your work as well as present it to the class.

Diffusion

Active transport

Facilitated

Diffusion

Osmosis

Water

Concentration

Proteins

Energy

Selectively permeable

For each type of molecular movement, remember to tell us

1. the direction that the molecules move, and

2. if it requires energy or something else.

FIGURE 16.2 Molecular movement rap (or poem or skit or story).

those who teach their children and to stay involved in the work that is brought home. If we wish to aspire to a culture of high expectations in school, then we must work together as a community to make it so. Rethinking homework and making it a community effort are necessary steps in the right direction and provide windows of opportunity for children to succeed in school.

REFERENCES

National Education Association. 2005. *Put an End to Homework Horror.* www.nea.org/classmanagement/ifc030202.html.

Pennsylvania Department of Education. 2004. Academic Standards for Science and Technology. www.pde.state.pa.us/k12/lib/k12/scitech.pdf.

17

Using Science to Bridge the Learning Gap between Home and School

Dale McCreedy and Jessica J. Luke

Trying to educate the young without help and support from home is akin to trying to rake leaves in a high wind.

—Wolfendale (1992)

Parent Partners in School Science (PPSS), developed by the Franklin Institute Science Museum, Philadelphia, was designed to use science teaching and learning to bridge the gap between parents and teachers in an urban school district. The program, with support from the National Science Foundation, targets elementary students, teachers, and families in three school communities. Museum educators work with teachers and parents to develop and coordinate inquiry-based science activities and events in the classroom, at home, and in the community. In this chapter, we discuss key lessons learned from the PPSS program, in order to share with both teachers and parents strategies for using science to support and integrate children's learning across home and school.

THE PROBLEM

Evidence and intuition suggest that the best educational experience for children is one in which the various influences in a child's life—home, school, and community—come together cohesively to promote achievement in a supportive learning environment. Anne Henderson and Nancy Berla (1994) reported that a child's academic achievement is directly related to the level of family involvement in her or his education, specifically the extent to which a child's family is able to create a home environment to encourage learning, express realistically high expectations for achievement and future careers, and get involved in their children's education in school and community contexts. Thus, the significant adults in children's lives play critical roles, not only in enriching their daily lives but also in shaping a future that includes reaching one's fullest potential.

Despite the mounting evidence that collaboration and communication between parents and teachers increase student achievement, however, it is still relatively rare for teachers and parents to bridge the communication gap between these two important influences in a child's life. The situation in urban areas—where diverse languages, ethnicities, socioeconomic and educational levels, and experiences prevail—only further complicates efforts to improve communication. Parents usually give several reasons for their lack of active involvement, ranging from economic and time constraints to linguistic and cultural differences. They often remark that they feel unwelcome in the school. In fact, studies have shown that, despite the rhetoric of promoting parent involvement in their child's education, many schools still—both intentionally and unintentionally—relegate parents and parent groups to marginal roles such as bake sales and signing forms and report cards. And, perhaps because they are given little, if any, training in strategies to encourage parental involvement, many teachers and administrators persist in believing that most parents cannot, or do not want to, play a more active role in their children's education. The clear dichotomy is that "most parents feel their concerns are given little consideration by educators, even though many teachers would like more parents to become engaged in their children's education" (American Association for the Advancement of Science 1996, 13).

RETHINKING PARENT INVOLVEMENT

In 1994, the U.S. Congress amended the national education goals and issued a much needed challenge to the nation's schools by demanding that schools "promote partnerships that will increase parental involvement and participation in promoting the social, emotional, and academic growth of children." In response, some schools have begun to make a concerted effort to solicit parental involvement. In fact, the National Parent Teacher Association released its *National Standards for Parent/Family Involvement* in 1997 (National PTA 1997), which provides voluntary guidelines for schools and parents to work together to effectively promote student achievement. The standards identify communication, advocacy, collaboration, and acknowledgement of the integral role parents play in their children's education as the foundation for programs that enhance parent involvement.

But what does the term *parent involvement* really mean? Certainly making a school more welcoming for parents, and inviting them to contribute their time in the classroom as volunteers, are strategies that many schools have begun to use in encouraging parents to become part of the school community. Recently, however, researchers from Columbia University's Teachers' College, and their colleagues, have raised some important questions about current definitions of parental involvement. Specifically, these researchers have used experiences in urban settings to challenge more traditional views of parent involvement. These views focus on the activities in which parents are involved, and how those activities fit or do not fit with the school's agenda. Instead, they recommend a shift away from the focus solely on *what* parents do to engage with their children's school to a focus on *how* and *why* parents are engaged, and the complex *ways* in which their engagement occurs. This new approach is termed *parent engagement* in a 2004 study reported by Angela Calabrese Barton, Corey Drake, Gustavo Perez, Kathleen St. Louis, and Magnia George.

Although the distinction may seem subtle, this new way of conceptualizing parents' engagement in schools allows us to recognize nontraditional ways in which parents can act and think about their child's learning. It places emphasis on two important aspects of parent engagement—capital and space. Capital represents all the various resources that parents bring to a situation, including not just knowledge and experience, but also beliefs and actions. Space acknowledges the many different physical and conceptual spaces in which parent engagement occurs, including school-based academic spaces as well as home-based or community-based spaces.

In our efforts to understand the successes and challenges within PPSS, we used this parent engagement framework to document the ways in which the program brought teachers and parents together in support of children's science learning. The framework was compelling to us because it helped us see that parent engagement is a dynamic, relational phenomenon, and that we should be trying not just to get parents involved in schools, but also to get parents and teachers to engage with one another across home and school.

A MODEL FOR USING SCIENCE
TO ENCOURAGE PARENT ENGAGEMENT

Working with three urban elementary schools in Philadelphia, PPSS was designed to foster home-school connections in support of K–4 students' science learning. Teachers, parents, and children were provided multiple opportunities for engaging in science activities and events at school, at home, and in the community. Specifically, the program was intended to accomplish three broad goals:

- Promote science teaching at the elementary level.
- Cultivate home-school collaboration in support of students' science learning.
- Document the role that a science center can play in bridging the gap between home and school in an urban educational district.

Specific programming efforts varied during the life of the project, as museum staff worked to customize activities and events to the realities and needs of each school. At a broad level, PPSS consisted of two main programmatic strands designed to foster home-school connections. The first strand targeted teachers, encouraging buy-in and ownership of the project within schools and providing teachers with resources and strategies for enhancing their science teaching. Key project components included Site Team Retreats (annual planning meetings for site team teachers, administrators, and parents), as well as Professional Development Workshops (annual workshops facilitating teachers' understanding of developmentally appropriate science content and teaching strategies).

A second strand of PPSS programming targeted families, encouraging parent engagement in schools generally and in children's science learning specifically. Major project components included Museum Adventure Days (open-ended exploration time for families in the museum), Discovery Days (theme-based workshops held at the schools, and designed to engage parents and children in the collaborative exploration of science), and Exploration Cards (home-based activities designed to offer nonthreatening opportunities for families to engage in science together).

Returning to the conceptualization of parent engagement discussed earlier, we focused our efforts on creating program components that would enhance both capital and space. In terms of capital, we wanted to communicate new messages to teachers and parents about what science is, and who can do it. Exploration Cards turned out to be an ideal way to do this. In terms of space, we wanted to provide multiple opportunities for parents and teachers to come together, to form new relationships, and, for parents in particular, to subtly reposition themselves within their child's school. Both Legacy Projects and Discovery Days accomplished this goal.

BUILDING PARENTS' AND TEACHERS' SCIENCE CAPITAL

Central to the PPSS project were efforts to enhance teachers' strategies for teaching science and their comfort level in teaching science. Professional development seminars accomplished this goal early on, but what really helped exemplify these teaching strategies was the development of Legacy Projects within each of the schools. *Legacy Projects*—a space or effort that could provide an opportunity or context for parents, teachers, and students to engage collaboratively in science learning—were initially conceptualized as a way to promote the development of an ongoing school-based commitment to science learning. Three unique spaces were created:

- A Science Discovery Room in an empty classroom, cleaned and painted by parents and teachers. The room contains activity boxes that have been developed by teachers and the Franklin Institute, piloted with parents during school visit days, and offered to families on weekend Discovery Days and as a resource to classes during the school day.

- A Science Garden, once a small plot of grass that now houses vegetable boxes and a butterfly garden. This space has been host to several tree plantings and Discovery Days, during which families and teachers plant bulbs, dig holes for trees, engage in related science activities, and learn about plants from an additional community partner, the Pennsylvania Horticultural Society.

- Stewardship of a corner of Philadelphia's Fairmount Park that provides a resource and shared physical connection for this one elementary school, housed in three separate sites. In partnership with the school, the Fairmount Park Commission, as well as the Franklin Institute, are working closely together to develop science programming for this natural space, both for weekdays and weekends.

Teachers at all three schools reported that they taught science more frequently, and felt more comfortable teaching it after participating in PPSS. In the words of one teacher,

> I really feel PPSS has [influenced my teaching] but it's hard to verbalize [how]. It's allowed me to delve into science more. It's encouraged me to see the daily connections to science, which in turn has helped me to encourage the kids to make daily connections. I feel more comfortable now. I used to be intimidated by science—it's so complex. But now I can see the simplicity of it, and help my kids see that too.

Parents also reported feeling more comfortable with science as a result of PPSS, claiming that PPSS helped them to better understand their child's interest in science, and gave them strategies for engaging their child in science at home. While parents also felt that the Legacy Project enhanced their child's interest in and perceptions of science, most parents pointed to the Exploration Cards as the primary source of their increased science capital. The Exploration Cards, intentionally created to foster nonthreatening, open-ended connections between in-school science learning and home-based experiences, require parent-child collaboration. Some cards have materials attached to them, while others are just paper-based. Upon completion, they are sent back to the classroom where they are then shared, and the content reinforced.

The following quotes indicate the positive response from PPSS parents:

> Usually with the schoolwork, there's so much that I'm like a sergeant—"Get that done!"—but with the Exploration Cards, they were smiling as we did them. I even got the little one involved in the activity. It was homework, but it didn't feel like homework. We were going about our daily stuff.

> [My children are] interested in things I didn't know they were interested in. I didn't know my daughter is interested in bugs. On paper they're okay. But she likes the whole discovering stuff. My son wants to be a fireman. This [project]

has enhanced his wanting to be in nature, helping people. Seeing how things work, helping people, how our actions affect outcome. And he's more adamant about what he wants. [Through PPSS,] I learned that my kids are smart. Not that I thought they were dumb, but it showed me how smart they are. They come home and tell me what they need to do.

I would say we definitely talk more now about it [science]. There is always an activity coming home that we do together, or they want to go to the museum and we're talking about science there. They are really reading more and more about science now, due to some of the free books they got.

CREATING SPACES FOR PARENTS AND TEACHERS TO INTERACT

PPSS also worked to shift the dynamic between parents and teachers, helping to break down some of the physical and psychological barriers to home-school connections. As a result of participating in PPSS, teachers expressed a stronger desire to work with parents, reporting that they communicated with parents more, facilitated by a common language offered through PPSS and an experience base through which to relate. Additionally, they felt less reluctant to invite parents into their classroom. Central to achieving these results were school-based events that were held throughout the year, and often within the Legacy Spaces. These events provided parents and teachers the opportunity to come together outside of the child's classroom, and to relate to one another in new ways:

> I think PPSS helped me to build better relationships with the parents in my classroom, so that when I needed to go to them about different things, anything, whether it be about the child's behavior in class or their child's academics, they weren't as intimidated by me. You know, because I'm the child's teacher and sometimes parents have a hard time because they see teachers as authority figures. And sometimes they don't want to deal with that. But I think PPSS has changed that environment.

> I've been able to use PPSS as a starting point in conversation with some parents that do not feel comfortable talking with teachers. I then am able to refocus the discussion to other areas, such as academics or behavior. I am interacting with parents more now because PPSS's philosophy is being generalized to other aspects of school life and we are, as a school, doing more to involve parents in our classroom and school activities.

Similarly, parents felt more welcome in their child's school as a result of PPSS. Parents had more positive feelings about their child's school, a better understanding of what their child was actually learning in school, and, perhaps most importantly, a different relationship with their child's teacher. Through PPSS, parents came to see their child's teacher in a new light, to see him or her as an "everyday person" with whom they could potentially converse, as opposed to an authority figure only to be spoken to during parent-teacher conferences. The following quotes typify these trends:

> You get welcomed into the school. You come into the school and as you're doing the activities, you get to see the school and the different classrooms. I got to talk to her science teacher for the first time, and see her science classroom for the first time. It opened up different aspects of school for me.

> I know more about the kind of science that they are learning. You know, they are learning so much at school, it's hard to know exactly what they're doing. But I feel like I know what science topics they are learning because of the cards and the events.

> I see her teachers at events, so I know them as more than just teachers. I get to see them not just at parent-teacher conferences, which is good because it makes the interactions more relaxed. They seem more like everyday people now.

CONCLUSION

Our experiences with PPSS, its challenges as well as its successes, have shaped each year in unique ways. We have learned a great deal about the ways in which school cultures differ, and have worked to realign and integrate program efforts into existing school cultures and structures in meaningful and effective ways. It is not surprising that community-based projects such as PPSS require multiple years, and repeated efforts, to demonstrate impacts. With the evolution of each of the project's events and components, we have come to better understand effective ways of engaging parents, encouraging science learning in the classroom and in the home, and building relationships among the various participants of PPSS, that is, the students, teachers, parents, and community, including the museum.

What is more, we have identified a conceptual model that encapsulates our thinking about how a community program like PPSS can bridge the gap between parents and teachers in an urban school district. By offering science programs in which parents, children, and teachers can all participate, and by offering nonthreatening and engaging ways for parents and children to engage in schoolwork together, we have created an entrée into science learning, and cultivated nontraditional ways of thinking about parents and schools. In other words, we have created space and capital for parents—many of whom have varying interests and skill levels—to begin to actively engage themselves in their children's formal education. Further, in cultivating nontraditional models of parent engagement, we have begun to expand teachers' ideas of ways in which they might engage parents. Through PPSS, science has begun to bridge the learning gap between home and school.

REFERENCES

American Association for the Advancement of Science (AAAS). 1996. *What Do Parents Need to Know to Get Involved?* NSF #MDR9550550. Washington, DC: American Association for the Advancement of Science.

Barton, Angela Calabrese, Corey Drake, Jose Gustavo Perez, Kathleen St. Louis, and Magnia George. 2004. Ecologies of Parental Engagement in Urban Education. *Educational Researcher* 33 (4): 3–12.

Henderson, Anne T., and Nancy Berla. 1994. *A New Generation of Evidence: The Family Is Critical to Student Achievement.* St. Louis, MO, and Flint, MI: Danforth Foundation and Mott (C. S.) Foundation.

National PTA. 1997. *National Standards for Parent/Family Involvement Programs.* Chicago: National PTA.

Wolfendale, Sheila. 1992. *Empowering Parents and Teachers: Working for Children.* London: Cassell.

ADDITIONAL RESOURCES

Suggested Readings

Epstein, Joyce L. 1996. Advances in Family, Community, and School Partnerships. *New Schools, New Communities* 12 (3): 5–13.

Epstein, Joyce, Lucretia Coates, Mavis G. Sanders, Beth S. Simon, and Karen Clark Salinas. 1998. *School, Family, and Community Partnerships: Your Handbook for Action.* Thousand Oaks, CA: Sage.

McCreedy, Dale, and Tobi Zemsky. 2002. *Girls at the Center: Girls and Adults Learning SCIENCE Together.* Philadelphia: Franklin Institute.

Websites

Center on School, Family, and Community Partnerships at Johns Hopkins University: www.csos.jhu.edu

Department of Education: www.ed.gov/parents/academic/help/tools-for-success/index.html

Harvard Family Research Project: www.gse.harvard.edu/~hfrp/

National PTA: www.pta.org

Parent Partners in School Science: www.fi.edu/ppss/

Research Links for Parents: www.ericfacility.net/ericdigests/ed419030.html

18

The Homeschooled Child and Science

Kathryn F. Cochran

Homeschooling is legal in all fifty states in the United States, but other than the fact that children are learning in a nontraditional situation, there is often very little similarity from one homeschool setting to another. There is a great deal of variety in nearly all aspects of homeschooling situations: in the types of instructional methods used, the curriculum materials, the amount of formal structure of the lessons, the legal requirements (which differ substantially from state to state), the reasons for homeschooling, the length of time learners spend in homeschooling (both daily and in terms of years), who makes the instructional decisions, and the subject matter content and assessment methods.

A FEW STATISTICS

To summarize the growth of this phenomenon, a few statistics might be useful. In 1994, the number of homeschooled children in the United States was estimated at 400,000. According to the National Center of Educational Statistics (2006), the number of children who were being homeschooled in 1999 was nearly 1 million, almost 2 percent of those between the ages of 5 and 17. In 2002, this number was estimated at between 1.7 and 2.1 million students, an annual growth rate of about 7 percent. The total is expected to reach 3 million students by the year 2010 (Jones and Gloeckner 2004). Broad support for this trend can be found in organizations such as the National Home Education Network, the American Homeschool Association, and the Homeschool Legal Defense Association (see websites listed at the end of this chapter).

REASONS FOR HOMESCHOOLING

Clearly this trend will continue. Moreover, the reasons parents give for homeschooling their children have become increasingly varied. While religious issues were initially a major impetus for the rise of homeschooling in the 1980s and 1990s (particularly for Christian groups), many parents are now making this decision for other reasons as well. The enormous amount of resources available on the Internet (see McDermott 2003) has virtually put the world at the fingertips of parents seeking teaching advice, curriculum materials, and classroom activities. The well-documented increasing diversity in the U.S. population has also fueled the homeschooling movement. Many general sources for homeschooling exist, but specific support can also be found centered in Native American, Jewish, Muslim, and African American philosophies and perspectives (for examples, see the end-of-chapter websites).

In addition, an increasing number of parents are questioning the rigidity and conformity required of children in traditional school settings, especially in the current political atmosphere of increasing accountability. Many parents question the amount and quality of useful and relevant learning that occurs in traditional settings. They may prefer their children to be exposed to a broader set of cultural views beyond traditional Western perspectives. Or, they may subscribe to nontraditional educational philosophies and methods. Advocates of homeschooling sometimes subscribe to a point of view known as *informal education*, where the goal is to focus on real-world, authentic learning experiences. There are also those who advocate *unschooling*, such as John Holt (2003), a pioneer of homeschooling, and believe that the main result of traditional schooling is to create competition and sort learners into hierarchies based on artificial criteria, a process that misdirects the advancement of society and global human cultural understanding.

COMMON MISCONCEPTIONS ABOUT HOMESCHOOLING

There are several common misconceptions about homeschooling and children who are homeschooled.

The most common misconception is that the social development of homeschooled learners is slowed or damaged as a result of an increased amount of time spent at home. Research shows that this is *not* the case (Ray 2000). Most parents who take on the responsibility of homeschooling do it with considerable thought and preparation, and often a family will make the commitment to live on one income instead of two in order to provide homeschooling for their children. These parents realize that their children need peer interaction and often specifically search out nonschool options such as lessons in dance or karate, or local organizations such as Boy Scouts, Girl Scouts, or community activities and sports programs. Travel is often considered a viable and valuable part of the homeschooling process.

A second common misconception is that children who are homeschooled have fewer opportunities for higher education. While homeschooled college applicants may have been at a disadvantage ten or fifteen years ago, an increasing number of colleges and universities are revising admission requirements to take this group of students into account (Jones and Gloeckner 2004). In addition, homeschooled college applicants are eligible for federal financial aid (Callaway 2004).

A third misconception is that children are homeschooled because they "can't make it" in traditional school settings. While this might be the case, the real issue is *why* they "can't make it." Homeschooled students are often those who would normally be classified as gifted students or special education students. In the case of special needs children, sometimes a school cannot afford the technology necessary to meet the needs of a specific learner; and sometimes the politics or policies of the school or system creates an extreme level of inflexibility. Overwhelmed administrators are sometimes unwilling to modify existing practices. Parents are often frustrated and disillusioned by these factors and turn to homeschooling. This is especially the case in nonurban areas, where choices of schools are limited (or nonexistent) or when private school costs are prohibitive.

With respect to academic ability, the research shows that homeschooled children score above average on college admission tests such as the ACT (Ray 2000). Several factors probably contribute to this result. First, one of the most solid findings in the literature is that one-on-one instruction yields high levels of learning, and homeschool settings routinely include very low, if not one-to-one, teacher-to-student ratios. Second, although research in this area is still limited, students appear to make more personal connections with subject matter, at least in science (Solomon 2003). Third, the segment of the population choosing homeschooling tends to be families that value both schooling and learning and have high expectations for children (Ray 2000).

A fourth major misconception is that homeschooled children are psychologically abnormal. There are some personality and temperament traits that do compel parents to homeschool their children, such as high levels of emotional awareness and empathy, shyness, unusual levels of maturity, or unusual susceptibility to stress or criticism—a common characteristic of gifted children. However, these traits are not abnormal, but are simply a less frequently occurring part of the normal continuum of individual differences.

For example, Elaine Aron (1996) writes about people who are highly sensitive and extremely aware of themselves and their environments. Although the majority of people are sensitive to some extent, for nearly 20 percent of the population this capability is advanced, and these individuals often feel overwhelmed by their surroundings. These are people for whom everyday life is as intense as being in a movie theater with the sound turned up. Traditional school settings, especially middle school settings, are not pleasant or productive places for learning for children who fall into this category. Children sometimes deal with the situation by focusing on coping with the social rather than the learning aspects of school, which can put them at risk of academic failure. The long-term consequences of such failure can be devastating. While the dominant society often dictates that parents teach their children to "be tough" and that "conflict builds character," this approach does not apply to all children and is inconsistent with and even contrary to many personal or cultural perspectives. (See the Chapters in this handbook under Part 4: "Equity.")

THE APA LEARNER-CENTERED PRINCIPLES

Some people have argued that parents generally do not have the science or the pedagogical background to homeschool children. However, just as there is increasing access to science information on the Internet, there is also increasing access to pedagogical and psychological knowledge on the Internet. An example of this is a set of fourteen basic learning principles created under the auspices of the American Psychological Association (APA) in 1997. These principles not only summarize a large body of research on learning but also highlight the importance of focusing on the learner and the needs of the learner, which is more easily done in a homeschooling setting. Research on learner-centered instruction is still in its infancy, but this approach is beginning to show increasingly positive effects on achievement.

HOW DOES THIS ALL RELATE TO SCIENCE?

First, science is an area where there are still documented "achievement gaps" between those learners who fit into the current culture of schools and those who do not. Many educators and parents believe that the slow pace of school reform justifies pursuing alternative schooling options, and that we should not be limiting the potential science achievement levels of any students in an era of rapidly increasing technological complexity. Some argue to the contrary: that alternative schooling options release the pressure on traditional school systems to adequately and fairly address the needs of all learners, and therefore accentuate the problems.

Second, the understanding of science can be directly related to the everyday world, and simple materials are easily accessible. What better place than the kitchen to "do" science? While I first had difficulty locating good science activities in the early 1990s for my own daughter, the number of resources has skyrocketed since that time. For example, there are at least five websites that describe how to extract DNA from an onion. Moreover, the simplest demonstration can spark great interest in a learner, and a homeschooling setting allows for the pursuing of that interest. This is a major factor in the enhancement of motivation that is internally rather than externally created.

Third, there are innumerable careers where science knowledge is used and accessible. Homeschooled children and their instructors can talk to and learn from pharmacists, beauticians, butchers, electricians, carpenters, farmers, chefs, pilots, bus and truck drivers, florists, artists, people at hobby stores and hardware stores, automobile mechanics, plumbers, zookeepers, grocery store produce managers, nurses and doctors, and so on.

A CASE OF HOMESCHOOLING

Having successfully negotiated traditional school settings through completion of a Ph.D. in educational psychology, I was hesitant to pull my twelve-year-old daughter out of school. I had the same misconceptions about homeschooling that I have outlined above. A gifted and very perceptive learner, she was about to embark on middle school. The day she said to me, "I don't care about learning any more, I just want to make it through school," was the day I realized that we had to make a choice between *learning in school* and *coping with school*. We chose learning.

Initially I thought I could easily create curriculum materials based on my own professional expertise and my background in science and mathematics. What I came to realize (to my surprise) was that this process was a full-time job. Fortunately that first year I was on sabbatical leave and had enough flexibility to manage. The following year, we enrolled in a private correspondence school, and my daughter's learning was much more successful. I am also certainly a better educational psychologist and advocate for children and their learning due to this experience.

Below is a set of quotations that are among those I have collected from parents and children about homeschooling over the past several years. They reveal some of the complexity of the processes of learning and schooling that we all still struggle with.

From children:

"I want to learn stuff, not go to school."

"I never thought of learning as making sense of something."

"Some teachers throw their knowledge at you."

"But what does it mean for an electron to be negative?"

"You mean I can make up my own math problems?"

From parents:

"Why should I let the school socialize my daughter into being a second-class citizen?"

"Why should I have to put my kid on Prozac to get him through middle school?"

"Why should my child have to spend a whole year learning to be a ninth grader when she is already more mature than that?"

"Why did my child's teacher tell her she wasn't good in science when she is—and more important, knows that she is?"

"The doctor told me that my child needed counseling because he wasn't adjusting to middle school."

REFERENCES

American Psychological Association (APA). 1997. *Learner-Centered Psychological Principles: A Framework for School Redesign and Reform.* www.apa.org/ed/lcp.html.

Aron, Elaine. 1996. *The Highly Sensitive Person: How to Survive when the World Overwhelms You.* New York: Carol Publishing.

Callaway, Sean. 2004. Unintended Admission Consequence of Federal Aid for Homeschoolers. *Journal of College Admission* (185): 22–28.

Holt, John C. 2003. *Teach Your Own.* Pasadena, MD: Holt Associates.

Jones, Paul, and Gene Gloeckner. 2004. Perceptions of and Attitudes toward Homeschool Students. *Journal of College Admission* (185): 12–21.

McDermott, Irene E. 2003. Web Resources for Home-Schooling. *Searcher* 11 (8): 27–31.

National Center of Educational Statistics. 2006. *Homeschooling in the United States: 2003.* http://nces.ed.gov/pubs2006/homeschool/estimated.asp.

Ray, Brian D. 2000. Home Schooling: The Ameliorator of Negative Influences on Learning? *Peabody Journal of Education* 75 (1–2): 71–106.

Solomon, Joan. 2003. Home-School Learning of Science: The Culture of Homes, and Pupils' Difficult Border Crossing. *Journal of Research in Science Teaching* 40:219–33.

WEBSITES

American Homeschool Association (AHA): www.americanhomeschool.org

American Psychological Association Learner Centered Principles: www.apa.org/ed/lcp.html

Bnos Henya Project: Jewish Orthodox Homeschooling: http://BnosHenya.org

Home School Legal Defense Association: www.hslda.org

Informal Education Network: www.infed.org/index.htm

Magazine for Muslim Home Schoolers: http://home.ici.net/~taadah/ula.html

National Center for Education Statistics (NCES): http://nces.ed.gov

National Home Education Network (NHEN): www.nhen.org

National Home Education Research Institute (NHERI), Salem, Oregon: www.nheri.org

Native American Educational Resources: http://lone-eagles.com/na-ed.htm

PART 4

EQUITY

19

Race, Equity, and the Teaching of Science

Garrett Albert Duncan

From its inception in the Western world as a formalized, disciplined system of inquiry, science has actively participated in constructing groups of people as different. For example, Carolus Linnaeus (1707–1778) catalogued Europeans and Africans in the influential *Systema Naturae* (1770) in the following manner:

> *European.* White, sanguine, brawny. *Hair* abundantly flowing. *Eyes* blue. *Gentle,* acute, inventive. *Covered* with close vestments. *Governed* by customs.

> *African.* Black, phlegmatic, relaxed. *Hair* black, frizzled. *Skin* silky. *Nose,* flat. *Lips* tumid. *Women's* bosom a matter of modesty, *Breasts* give milk abundantly. *Crafty,* indolent, negligent. *Anoints* himself with grease. Governed by caprice. (Italics in the original.)

In *On the Natural Varieties of Mankind* (1775), J. F. Blumenbach (1752–1840), a disciple of Linnaeus, would later divide humankind into the all-too-familiar color categories that he called "races." Scientists would later reify this concept during the nineteenth century and, in doing so, endow the idea of race with explanatory powers far beyond the hints of ranking evident in Linnaeus's approach to classification. Indeed, they imbued the concept with more meanings than what the relatively egalitarian Blumenbach demonstrated across his writings. For instance, science went on to support a global system of white supremacy, a brand of racial hierarchy that was evident in the colonial expansion of Europe across Asia, Africa, and the Americas as well as in the various structures of white dominance in the United States (e.g., slavery, Manifest Destiny, and segregation).

RACE AND SCIENCE IN NINETEENTH- AND TWENTIETH-CENTURY AMERICA

Ironically, during the periods in the United States when what is now considered racist research was being conducted by some of the most distinguished scholars of the day, black people made basic contributions to the various, often fledgling, sciences. For example, while "scientific racism" was still in its adolescence, Benjamin Banneker (1731–1806), the "Sable Astronomer," had already regularly published an almanac. Comprised of information tables that forecasted meteorological events, the annual publication was a best seller from 1792 to 1797 in Pennsylvania and Virginia. In 1867 Edward Alexander Bouchet (1852–1918) became the first black recipient of the doctorate, a Ph.D. in physics from Yale, and the first black member of Phi Beta Kappa. Later, during a period when "scientific racism" was still regarded as legitimate in the academy and respectable in the wider society, W.E.B. Du Bois (1868–1963) became one of the founders of the field of sociology, Sarah Breedlove McWilliams Walker, aka Madame C. J. Walker (1867–1919), had invented medicinal treatments to heal scalp disease and was on her way to becoming an innovative and prosperous entrepreneur; and Ernest Everett Just (1883–1941) made fundamental contributions in the area of the physiology of development.

Despite the apparent contradictions in society that belied the existence of inherent moral and intellectual differences between the races, law and custom—informed by white supremacy and buttressed by science—severely limited the access that racialized U.S. cultures had during the late nineteenth and early twentieth centuries to education. These groups encountered their greatest impediments in their pursuit of education in the fields of biology, mathematics, and the physical and the earth and planetary sciences, especially at the most advanced levels. For example, between 1876 and 1943, there were 119 doctorates awarded to African Americans in the aforementioned fields. In contrast, during the same period, there were more than 3,000 doctorates awarded to African Americans in education and psychology (Jay 1977). The number of doctorates awarded in the biological and physical sciences to African Americans almost quadrupled over the next quarter of a century. However, the numbers relative to their general population in society remained disproportionately low for African Americans at the dawn of the civil rights movement of the mid-1950s and early 1960s; such conditions held true also for other racialized U.S. cultures.

RACE, SCIENCE, AND CIVIL RIGHTS

The civil rights movement, revived by mid-century domestic and international exigencies, coincided with the scholarly refutation of scientific racism.

According to Elazar Barkan (1992), the atrocities that occurred in Europe between the two world wars contributed to the increased presence of Jewish intellectuals, as well as of women and leftist scholars, in British and American colleges and universities. The presence of these scholars provided an impetus for a changing conception of race in the academy, where racial differences were no longer being attributed to biological but to cultural differences. In addition to being influenced by political events outside of the academy, scientists began to assert their influence on social and public issues beyond the ivory tower, especially in the area of race relations. Perhaps the strongest statement of this kind came in 1950 in the United Nations Educational, Scientific and Cultural Organization's (UNESCO) "Statement on Race," in which scientists recognized "that mankind is one: that all men belong to the same species, *Homo sapiens*" (496).

Science also played a prominent role in removing the barriers that prevented racialized U.S. cultural groups from gaining access to quality education, as was evident in the role of the social sciences in the U.S. Supreme Court's ruling in the landmark *Brown v. Board of Education* (1954). This court decision led to the slow but substantial integration of elementary and secondary public schools, especially in the South, as well as in the dramatic increase in the number of students of color in predominately white universities and colleges by the 1970s and the 1980s. The increase in the number of students of color in American colleges and universities resulted in the eruption of ethnic studies programs in American colleges and universities. Significantly, professors in the behavioral and social sciences, especially psychology, founded a number of these programs. In addition, the increase in the number of minority students on college campuses was the impetus behind the establishment of organizations and programs to support these students as well as to recruit and prepare those in precollege settings for success in higher education, especially in the areas of science and engineering. In 1971, for instance, two undergraduates at Purdue University founded the Society of Black Engineers, now the National Society for Black Engineers (NSBE), to improve the recruitment and retention of black students in the field. NSBE now has a membership of 15,000 members, 17 precollege programs, and 268 student and 50 alumni/technical professional chapters.

In 1973 a group of science professors, professional scientists, K–12 educators, and students founded the Society for the Advancement of Chicanos and Native Americans in Science (SACNAS). SACNAS's mission is to increase the number and the quality of the experiences of Chicano/Latino and Native American students who pursue graduate studies in science and engineering. An additional goal of the organization is to increase the number of these students who obtain the advanced degrees necessary for research careers and for science teaching at all levels. Similarly, in 1974 a group of professional engineers founded the Society of Hispanic Professional Engineers (SHPE) in Los Angeles and soon established chapters across the nation, eventually becoming international in scope. School-university partnerships established to increase the number of students of color in the sciences include the Southeast Consortium of Minorities in Engineering (SECME), the Mathematics and Science Education Networks (MSEN) in North Carolina, and the Mathematics, Engineering, Science Achievement (MESA) programs, which are largely concentrated in California. In addition, civil rights organizations, such as the NAACP, through its Academic, Cultural, Technological, and Scientific Olympics (ACT-SO), have also promoted science education among traditionally underserved youth populations.

SCIENCE EDUCATION IN CONTEMPORARY SOCIETY: A RACIAL REPORT CARD

The scientific renaissance of the 1970s and the mid-1980s was given a boost in 1989 with George H. W. Bush's Goals 2000 initiative, in which he pledged to make America's children and youth "first in math and science." Indeed, by the 1990s African American, Chicano/Latino, and Native American students had gained greater access to education in the sciences at all levels and had increased the percentage of degrees in the sciences they earned relative to their population in the general society. However, a number of disparities still remain among ethnic and racial groups in precollege, undergraduate, and graduate education, resulting in continued inequalities in the broader society at large.

ELEMENTARY AND SECONDARY EDUCATION

For example, according to the National Center for Education Statistics (NCES), the 2000 National Assessment of Educational Progress (NAEP), or the "nation's report card," indicates the existence of disparities among racial and ethnic groups in the area of mathematics achievement in elementary, middle, and high schools (Table 19.1). Achievement disparities in mathe-

TABLE 19.1 NAEP Average Mathematics Scores for Public School Students by Race/Ethnicity and Grade, 2000

Race/Ethnicity	Grade 4	Grade 8	Grade 12
White	235	285	307
Black	205	246	273
Hispanic	211	252	281
Asian American/Pacific Islander	–	288	318
American Indian/Alaskan Native	215	261	292

Source: National Center for Education Statistics, National Assessment of Educational Progress (NAEP), 2000 Mathematics Assessment.

TABLE 19.2   NAEP National Mathematics Results for Grade 8: Public School Percentages of Students below, at, or above Achievement Levels by Race/Ethnicity, 2000

Race/Ethnicity	N	Average Scale Score	Achievement Level			
			Advanced (%)	Proficient (%)	Basic (%)	Below Basic (%)
White	4,988	285	6	34	77	23
Black	1,854	246	0	5	32	68
Hispanic	1,909	252	1	9	40	60
Asian American/ Pacific Islander	451	288	11	40	75	25
American Indian	133	261	0	12	50	50

Source: National Center for Education Statistics, National Assessment of Educational Progress (NAEP), 2000 Mathematics Assessment.

matics are brought into even greater relief when viewed in terms of how these scores are distributed within each group across achievement levels. Here, two-thirds of black and more than half of Latino and Native American eighth graders are performing below basic levels of achievement compared to a quarter of their white and Asian American peers (Table 19.2). Similar disparities are evident in the area of science. For example, according to the NAEP 1996 science results, three-quarters of black, two-thirds of Latino, nearly half of Native American, and more than a third of Asian American students

TABLE 19.3   NAEP National Mathematics Results for Grade 8: Weighted Percentages of Students below, at, or above Achievement Levels by Race/Ethnicity, 1996

Race/Ethnicity	N	Weighted Percentage	Achievement Level			
			Advanced (%)	Proficient (%)	Basic (%)	Below Basic (%)
Total	7,774	100	3.0	28.9	61.3	38.7
White	4,292	69.6	4.1	36.9	73.3	26.7
Black	1,492	14.1	0.2	5.0	24.0	76.0
Hispanic	1,426	11.8	0.3	10.7	35.9	64.1
Asian American/ Pacific Islander	382	2.6	3.0	30.5	61.9	38.1
American Indian	149	1.7	2.5	24.2	59.5	40.5

Source: National Center for Education Statistics, National Assessment of Educational Progress (NAEP), 1996.

TABLE 19.4   NAEP National Mathematics Results for Grade 12: Weighted Percentages of Students below, at, or above Achievement Levels by Race/Ethnicity, 1996

Race/Ethnicity	N	Weighted Percentage	Achievement Level			
			Advanced (%)	Proficient (%)	Basic (%)	Below Basic (%)
Total	7,537	100	2.6	21.2	57.1	42.9
White	4,748	69.9	3.4	27.1	68.0	32.0
Black	1,225	14.3	0.1	3.7	23.3	76.7
Hispanic	1,015	11.2	0.8	7.1	33.2	66.8
Asian American/ Pacific Islander	458	3.8	3.0	21.9	55.8	44.2
American Indian	70	0.7	0.5	10.5	51.9	48.1

Source: National Center for Education Statistics, National Assessment of Educational Progress (NAEP), 1996.

performed below basic achievement levels in contrast to a quarter of white students who did so (Table 19.3). The percentages in the *below basic* category actually increased for all groups for grade 12, with even greater disparities indicated between white students and their Asian American and Native American peers (Table 19.4).

These racial disparities mirror significant differences in the way teachers instruct students, especially when it comes to the use of computer technologies. For example, in 1998 more teachers reported using computers primarily for drill and practice with their black eighth-grade students (42 percent) than they did with their white (35 percent), Asian American (35 percent), or Latino (35 percent) eighth graders. In contrast, fewer of these teachers reported simulations and applications or learning games as their primary computer use with black students (14 and 48 percent, respectively) than they did with their white (31 and 57 percent), Asian American (43 and 57 percent), and Latino (25 and 56 percent) students (Wenglinsky 1998). These findings reflect a timeworn pattern in which teachers routinely employ qualitatively different curricular and instructional strategies with their students in ways that place black students at a disadvantage in higher education and in a high-tech socioeconomic order that requires of its participants innovation, creativity, intellectual dexterity, and initiative.

UNDERGRADUATE AND GRADUATE EDUCATION

Similar disparities related to race and science exist at the undergraduate and graduate levels in terms of the degrees that are distributed to students, relative both to their proportion in the general population as well as to comparisons among racial and ethnic groups. The U.S. Census Bureau estimates that in 2001, about 62 percent of traditional college-age students (ages 18–24) were white, 17 percent were Latino, 14 percent were black, 4 percent were Asian American, and just less than 1 percent (0.9 percent) was Native American. Of the 27,000 doctorates awarded to U.S. citizens in 2001, 15,000 (56 percent) were granted in the fields that encompass the sciences and engineering (Division of Science Resources Statistics 2001). Of these, 80 percent (12,211) were granted to white candidates. Seven percent (1,058) of the science and engineering doctorates were awarded to Asian American students, with approximately 4 percent granted to African American (625) and Latino (581) candidates and 0.5 percent (72) given to Native American candidates. There were discernible racial differences with how these doctorates were distributed. One-third of the doctorates awarded to Native Americans (37) and one-fifth of those awarded to Latino (253), African American (334), and white (2,449) recipients of the Ph.D. were in the behavioral and social sciences; in contrast, about 11 percent (221) of Asian Americans who received doctorates in 2001 did so in these areas. Biological sciences and engineering comprised the largest block of nonbehavioral/social science doctorates awarded in 2001. This includes 23 percent of white doctoral recipients (3,124 biological and 1,645 engineering) compared to 46 percent of their Asian American (365 and 261), 24 percent of their Latino (149 and 73), 14 percent of their black (121 and 82), and 18 percent of their Native American (15 and 17) peers who also received advanced degrees in the biological sciences and engineering that year.

In contrast to the advanced degrees distributed in the aforementioned areas, U.S.-born citizens of all backgrounds were granted less than half of the doctorates awarded in either the computer sciences (365 of 826) or mathematics and statistics (468 of 1,006). Of these, 83 percent were awarded to white candidates (298 computer science, and 395 mathematics and statistics). Asian American candidates received more than half of the remaining doctorates in the computer sciences (9 percent, or 34), followed by black (3.6 percent, or 13) and Latino (1.6 percent, or 6) Ph.D. recipients in these areas; no Native American received a doctorate in the computer sciences in 2001. Asian American candidates were awarded nearly half of the remaining doctorates in mathematics and statistics (6 percent, or 30) in 2001, followed by their black (3.6 percent, or 17), Latino (3.0 percent, or 14), and Native American (0.2 percent, or 1) peers.

Unfortunately, the disparities described in the previous section may not be reduced in the near future. According to a survey conducted by the Higher Education Research Institute (HERI) of the University of California, Los Angeles, approximately a third of black, Latino, and white and 40 percent of Asian American first-year students in the incoming class of 2002 indicated their plans to declare either science or engineering majors. Thirteen percent of all Latino students indicated their plans to major in the social and behavioral sciences, the top choice of this group among the science and engineering

TABLE 19.5   Intentions of First-Year Students at All Four-Year Colleges to Major in Science and Engineering Fields, by Race/Ethnicity, 2002 (%)

Race/ Ethnicity	All Science and Engineering Majors (%)	Biological/ Agricultural Sciences (%)	Computer Sciences (%)	Engineering (%)	Math/ Statistics (%)	Physical Sciences (%)	Social/ Behavioral Sciences (%)
White	30.5	7.0	2.0	9.2	0.8	2.1	9.4
Black	34.2	8.4	3.5	8.6	0.5	1.2	12.0
Chicano/ Puerto Rican	34.1	8.2	3.5	8.6	0.7	1.6	13.2
Asian American	42.1	11.9	3.5	15.9	0.8	1.9	8.1
Native American	31.0	7.9	1.8	7.5	0.6	1.7	11.5

Source: Higher Education Research Institute, University of California, Los Angeles, Survey of the American Freshman, 2002.

majors. Similarly, 12 percent of all black and Native American students have chosen these majors as their favorites among the sciences, as do their white peers (9.4 percent of the entire college-attending members of this racial group). In contrast, first-year students, in general, indicate a far lower preference for majors in the computer and physical sciences, with even fewer students choosing to pursue mathematics degrees (Table 19.5). Given the implications that such preferences hold for the socioeconomic requirements of the twenty-first century, the sciences and mathematics are indeed the cornerstones of social justice for racialized U.S. cultures in the new millennium.

CONCLUSION: RACE AND SCIENCE—A ONCE AND FUTURE RELATIONSHIP?

The noted social theorist Anthony Giddens (2000) explains that the world in which we now live is much more complicated than the ones of the past because of the proliferation of media and other forms of technologies that allow for the unchecked dissemination of information, images, and symbols. This is a "runaway world," he notes—one that is associated with drastic social and economic changes in both the United States and the broader international society. These changes, attendant to postindustrialism and globalization, present individuals with a vast array of social, cultural, and economic opportunities. At the same time, opportunities have not been available to all, and, so far, both postindustrialism and globalization have reinforced patterns of racial dominance, both in the United States and abroad. For instance, jobs in the United States for the foreseeable future will be divided between disproportionately high numbers of opportunities in low-paying, low-status, unstable positions and small numbers of high-paying, high-status, more secure ones. Racialized U.S. cultures are generally confined to the former category and increasingly are being left out of both. Further, social and economic disparities in contemporary society tend to reinforce timeworn racial stereotypes. In addition, they have created the conditions for the proliferation of scientific studies that are eerily reminiscent of Old World race scholarship to provide pat explanations for social stratification.

Thus, changes attendant to postindustrialism and globalization have created a different set of imperatives for education, at all levels, for all students in the sciences and *especially* for students from racialized U.S. cultures. Science education not only positions these groups to access the resources to empower them to eliminate material conditions of inequality. It also provides them access to the means of production to challenge the print and visual media—the products of new century science—that figure so prominently in constructing the imagery that provides the commonsense understandings of, and even the justification for, their conditions.

REFERENCES

Barkan, Elazar. 1992. *The Retreat of Scientific Racism: Changing Concepts of Race in Britain and the United States between the World Wars.* Cambridge: Cambridge University Press.

Blumenbach, Johann Friedrich. 1969. *On the Natural Varieties of Mankind.* New York: Bergman Publishers. (Originally published in 1775.)

Division of Science Resources Statistics. www.nsf.gov/statistics/

Giddens, Anthony. 2000. *Runaway World: How Globalization Is Reshaping Our Lives.* New York: Routledge.

Jay, James M. 1977. *Negroes in the Sciences: Natural Science Doctorates, 1876–1969.* Detroit, MI: Balamp Publishing.

Linnaeus, Carolus. 2003. *Systema Naturae.* Tuurdijk: HES & DE GRAAF Publishers. (Originally published in 1735.)

National Center for Education Statistics. http://nces.ed.gov/.

UNESCO. 1950. *Race and Science.* New York: Columbia University Press.

Wenglinsky, Harold. 1998. *Does It Compute? The Relationship between Educational Technology and Student Achievement in Mathematics.* Princeton, NJ: Educational Testing Service.

ADDITIONAL RESOURCES

Suggested Readings

Gould, Stephen Jay. 1981. *The Mismeasure of Man.* New York: W. W. Norton.

Harding, Sandra, ed. 1993. *The "Racial" Economy of Science: Toward a Democratic Future.* Bloomington: Indiana University Press.

Herrnstein, Richard J., and Charles Murray. 1992. *The Bell Curve: Intelligence and Class Structure in American Life*. New York: New Press.
Rushton, Jean-Philippe. 1995. *Race, Evolution, and Behavior: A Life History Perspective*. New Brunswick, NJ: Transaction.

Websites

Higher Education Research Institute: www.gseis.ucla.edu/heri/heri.html
Mathematics and Science Education Network: www.unc.edu/depts/msen/ and www.ncsu.edu/crmse/programs/msen.html
Mathematics, Engineering, Science Achievement: http://mesa.ucop.edu
National Center for Education Statistics: http://nces.ed.gov
National Science Foundation: www.nsf.gov
National Society of Black Engineers: www.nsbe.org
Society for the Advancement of Chicanos and Native Americans in Science: www.sacnas.org
Society of Hispanic Professional Engineers: www.shpe.org

20

Exploring Issues of Race, Power, and Concern in Rural Lower-Track Science Contexts

Randy K. Yerrick

Antwann is an overweight young black man who daily sits at the back of yet another middle-aged white man's classroom. Complacency and disengagement are the tools of Antwann for minimizing conflict with teachers. He doesn't ask me for anything except an occasional request to visit the bathroom. I had asked the students to think about a question that was meant to simply solicit student thinking: "Where do you think clouds came from?"

> *Teacher:* Who would like to share next? How about you, Antwann? I am sorry I didn't get back there, and I want to make sure I get everyone's contributions before we move to finding out the answer.... Antwann?
>
> *Antwann:* What?
>
> *Teacher:* What were your ideas about where clouds came from?
>
> *Antwann:* Huh?
>
> *Teacher:* I want to know where you think clouds come from.
>
> *Antwann:* Oh, I don't know.
>
> *Teacher:* Well, what did you share with David over there?
>
> *Antwann:* Nothin'.
>
> *Teacher:* Why not?
>
> *Antwann:* We didn't talk or nothin'.
>
> *Teacher:* Okay then...um...well what did you write down?
>
> *Antwann:* Nothin'.
>
> *Teacher:* Really? Nothing at all? (*Long pause in my bewilderment.*) Well why not?
>
> *Antwann:* I don't know.

It was evident from his posture and facial expressions that he was growing more and more uncomfortable with being the center of attention, so I moved on with other students and discussed his participation with him later. I wondered just what kind of school was Antwann used to and why this level of disengagement seemed appropriate to him. How had this come to be known as a normal experience in school?

Because such issues with lower-track learners cannot be examined, much less changed, from an outsider's perspective, I volunteered to teach at Antwann's school, just a short drive from my university, for the year. Though I never expected to meet Antwann, his case became important for making connections for my preservice teachers preparing for the world. My characterization of this learning context is heavily influenced by Nel Noddings's (1992) notions of "caring" for students as I sought to find an intervention for Antwann at his school. In her book *The Challenge to Care in Schools*, Noddings argues that school structures, routine practices, and rewards present obstacles to educators who would make trusting, caring, and concern important tenets in their definition of success with students. Few authors seriously chronicle the challenges of assisting students who operate within the confines of science classrooms, but those few who have done it well understand the extent to which schools' structures and routine practices need to be rethought in order to result in the inclusive, equitable, and democratic science classroom visions that reformers continue to buffet practitioners with.

When the year began, it was clear that some students would have greater difficulty than others. Antwann was among them. I knew that Antwann was marked by the principals as trouble the first day they followed him all the way to my class to confront him about school policies and his wearing a "dew rag" in my class. Being new, I did not understand the implications, but I promised the principal that Antwann and I would both comply to this school policy. Administrative responses were swift and sure, and perhaps there were days he intended on contesting school rules. I continually wondered if Antwann was actually embracing the dangerous lifestyle depicted by his favored rap artists or whether it was just a ploy to have a center-identity. An attempt to be nonwhite in this rural, white-dominated, rural school context would have been a typical response (Gilbert and Yerrick 2001).

I decided that I needed to understand far more about Antwann to ever hope to engage him in science inquiry activities. As I learned more, I realized how much more enriched I was understanding the school and other cultures, but there was often a sense of overwhelming hopelessness caught up in my dealings with Antwann. I found myself often wishing that I had not asked or that I did not know certain tragic circumstances that had befallen this young man. I was caught between needing to know and wanting to forget about much of what I learned.

My first stop was the counselor's office to get a brief academic history of Antwann. It was somewhat expected—he had poor English skills, and had not passed the eighth-grade reading or math competency exam. More than two-thirds of the students enrolling in my earth science class had not passed their eighth-grade reading and mathematics competency tests—tests that were representative of only the most basic, objective, and factual treatment of received knowledge. Antwann had been labeled for years and had received some help from individual teachers, though I was informed he didn't entirely qualify for "services." His troubles were not "profound enough," I was told by one counselor, and subsequently Antwann had become one who had "fallen through the cracks."

Teachers and other administrators conveyed (sometimes unsolicited) deficit perspectives of students in discipline reports using phrases such as acts of insubordination, poor family background and no real family structure. Some more general disparaging descriptors like "laziness" were used more as factual personal attributes than subjective observations. One counselor's description was particularly disparaging.

> *Yerrick:* Can you tell me about Antwann? How did he get in this earth science?
>
> *Counselor:* Antwann...F...Bowden. Now there's a case! I am not sure if you have enough time for this story.... He's operating at just the *basic* level here, just trying to get enough credits to graduate. Let's see... *(scanning the computer screen)*...nineteen years old...fifth year here...hasn't passed the basic eighth-grade reading or mathematics competencies. I recommended that he take welding at the local community college as an elective that would help him develop a useful skill. He just needed something to get him through so that he could get outta here. Yeah, I don't envy ya there. You're really operating in the negative realm there. Don't know what to tell ya about Antwann.

But Antwann was nineteen and wanted out. Along with many of his peers, Antwann represented a group of students who typically choose not to engage in classrooms the way teachers or institutions expected them to. Frederick Newmann (1992) treated *alienation* and *engagement* as antonyms—characterizing alienation with terms such as *detachment, isolation, fragmentation, disconnectedness, estrangement,* and *powerlessness.* So, in effect, their lack of engagement in daily tasks is testament to their alienation. My students had learned their disengagement in school, and this made them even less accessible to teachers desiring to follow Noddings's (1992) challenge to care.

Despite the warnings I was given by counselors and advice that there was not much I could do, I thanked them (as one is expected to do in the rural South despite how distasteful the interaction) and decided to press beyond their grim prognosis. I took their advice in light of Ladson-Billings's descriptions of such deficit thinking as she wrote,

> We may want to believe that this different group of students requires some extraordinary type of teaching because if we do not believe it, it calls into question all the teaching we have endorsed heretofore. (Ladson-Billings 1999, 242)

I noted that Antwann was responding to my invitations to participate but did not know if he continued disengaging in other classes. How did his peers and their antisocial, antischool beliefs and behaviors affect Antwann outside of my class? I decided to shadow Antwann and find out what the rest of his school day was like. Shadowing students to gather information with their permission and cooperation to reconstruct their individual school experience was a practice I commonly assigned to my preservice teachers who regularly observed my lower-track class (Yerrick 2004). My assignment would be Antwann.

My first stop was the class right after mine. I tried to wrap up my class and get to Antwann's second hour before the bell rang, but as usual I was late in packing up my things, collecting papers, and erasing the board of the teacher who was so generous to let me borrow her room for first hour. As I left the classroom, I was reminded how different my students are as one black and twelve white students filed in quietly past me, sat at the front of the room, and eagerly opened their AP Biology notebooks for their lecture. When I entered Antwann's second hour late, I found Antwann sitting with the same peer group he would follow with the rest of the day, clearly displaying similar school dispositions. Some had their heads down, and others were staring a blank look out the windows. I explained to the teacher that I was his first-hour teacher and that I was observing to figure out how to assist Antwann in the rest of his studies. She obligingly let me watch during the class. What I was struck by immediately bordered on absurdity as I assessed the relevance of this activity to Antwann. For the first hour, an invited speaker for the day, a white woman in her late sixties, was discussing quilting projects and what she was making for Christmas presents. On and on she went without interruption and without noticing that five of the students had actually turned their chairs away from facing her and half the class had their eyes closed. I tried to make connections with Antwann like the clothes he was wearing, but I choked on the words *relevant* and *interest* as they just came out muddled. The second part of the class, she had them practice sewing straight lines with a sewing machine. Big Antwann dwarfed the machine, nearly enveloping it as his fat fingers labored to force strips of paper through the machine. After several attempts and rethreading the machine, Antwann finally emerged with a fairly straight line on his paper strip that he handed in for a grade. I was taken aback by the learning metaphor represented by this paper strip. They were not even being asked to learn to sew for any particular outcome or product beyond a paper strip that would surely be thrown away by the teacher after grading. What could be more temporary or fading than this representation of learning with such shallow and easily tattered purpose?

The crumpled paper in the trash basket seemed a perfect example of McNeil's (1986) description of content so disparate and foreign to this student's life and unrecognizable from the connected body of knowledge from which it had been derived. McNeil argued,

> Ironically their [teachers'] attempt to minimize student cynicism by simplifying content and avoiding class discussion only heightened student disbelief of school knowledge and fostered in students greater disengagement from the learning process. (1986, 160)

In other words, students engage with schools in ways best explained by continued skepticism and disillusionment with what schools can do for them, and with questions about the legitimacy of the role of school in their lives. While schools interpret misbehaving youth as a control problem, McNeil argued that skepticism and disillusionment are more fundamental reasons why students resist.

Antwann's third and fourth classes were study skills. There was no credit associated with them but simply a holding pattern for the two English classes he needed to graduate—taken later in the day. This, it seemed to me, was another less than benign school experience that suggested why Antwann found it so easy to disengage from my class. As I entered, I found only tables across the front and walled carols at the back. The room was crowded, with no visible windows, and was lit with painfully bright fluorescent lamps humming above. The room smelled of poorly kept bodies, strong Afro hair treatments, and sweat. Almost all of the students were black males and had attended this school for at least two years. Antwann was one of the first few to enter the class, and as I followed him in I noticed that he did not acknowledge the teacher across the room. He kept his head down, giving a nod to a fellow student as he sauntered to the back of the class.

I excused myself to go and speak with the teacher to explain my presence. She was a shorter, middle-aged white woman with a commanding voice. When I told her that I was here to help Antwann, at first she laughed—right out loud. Could it be that she didn't remember that I was a fellow teacher truly interested in Antwann's learning? I was quiet, and my presence in the school was a novelty to most of the faculty. Only a few teachers were good friends of mine, and it was hard to make new friends as an outside professor in just the first few weeks of school. Perhaps she thought I was a temporary social worker who didn't quite understand the situation and who had less than a vested interest in helping with Antwann's academic struggles. But her next response told me that it mattered little who I was. Her room was connected to that of a fellow teacher by way of an office space. Antwann's teacher shouted to her office mate, "He's here to help Antwann!" Her office mate must have been familiar with Antwann's academic history as well, for she responded with a rolling of her eyes and shared the chuckle.

I wasn't sure what was funny, and my heart broke with the abrasiveness of criticism of me and Antwann, made public for all the class to experience. No one else in the class seemed to take notice, though. I returned to the back of the classroom, where I found Antwann talking to a fellow student— undaunted by the previous encounter. The bell rang and many students hung around with others in the hallway, acting boisterously with pushing and yelling and a general bravado from both males and females.

My day continued, class after class, teacher after teacher, each being confronted by the same challenge. "How can I care for this older student telling me through his dress, words, and very actions he doesn't want to learn?" Antwann is not a passive spectator but often an active initiator of much of his disengagement. Like Willis's lads (1977), my students' actions were not those of acquiescent subjects of the process, but often those of bold initiators of their own alienation from the school. As in the case of Willis's lads, classrooms manifest class struggles in which teachers and students were primary players. As others have described (e.g., LeCompte and Dworkin 1991), this was a dynamic in which teachers and students were at direct odds.

So the challenge for teachers to care and the challenge for Antwann to believe there is good reason to engage in classroom activities raise significant questions about how to help students acquire new ways of speaking, thinking, and acting in science class. How can we simultaneously change the way lower-track students think about their investment in scientific concepts presented by a teacher while the student questions the very agency and purpose of the messenger? Are school, culture, and science teaching so inextricably tied that one cannot be transformed without totally unraveling the remainder of school fabric? Clearly such a transition from a factual-based constrained discourse to an open-ended construction of scientific evidence-based arguments required students to do significant code switching between home-based discourse, lower-track classroom discourse, and the new discourse of our evolving learning community.

Antwann and I continued to transition toward different roles, rules, and relationships implicit in each of the discourse communities he encountered in his day, but changing the classroom discourse was not a matter of simply replacing what did not work with something that would. We had to relearn and renegotiate the rules of participation in my classroom. I had to learn to listen carefully to the way my students talked and the origins of the mistrust they felt in a context that had treated them as outsiders. Together we worked to find ways to participate appropriately as members of a new community, but not without facing with great courage Noddings's (1992) challenge to care.

A primary goal of this chapter is to complicate the reader's potential bias toward Antwann or those he encounters during his day. It is not a one-sided problem. Antwann's engagement in school is not easily solved but is more aptly framed when we consider the challenge Noddings has given us to "care in school." Antwann makes it difficult to care for him, and his teachers make it difficult for Antwann to trust. It is a broken system from both ends, which has few easy solutions. Those solutions, however, are likely to emerge from more closely examining what students know and can do, their ways of participating in multiple settings, and the ways in which norms get renegotiated in classes. These are still undeveloped ideas that lie outside the norm or common language to understand teaching or learning of science but are sure to bear fruit in future studies undertaken by teacher researchers. I consider my experience with Antwann a vital one to my learning as a teacher and researcher that helped me see school in new ways. Delpit advised that when those with power attempt to understand the perspective of the powerless, *"suddenly many of the 'sensible' ways of doing things no longer seem acceptable"* (Delpit 1995, 73 emphasis in original). I believe that for science teachers to learn to combat the challenges to care in schools, we must often see and encourage our students to see as Delpit has advised—to see one another differently as a means to overcome critical barriers related to students' identity, alienation, disinterest, and skepticism. Real change is needed. Without thoughtful consideration of such interactions, efforts to change these problems through revised curricula, administrative policies, or even future research will be as futile a paper effort as Antwann's sewing assignment—inappropriate and soon forgotten.

REFERENCES

Delpit, Lisa D. 1995. *Other People's Children: Cultural Conflict in the Classroom*. New York: New Press.

Gilbert, Andrew, and Randy Yerrick. 2001. Same School, Separate Worlds: A Sociocultural Study of Identity, Resistance, and Negotiation in a Rural, Lower Track Science Classroom. *Journal of Research in Science Teaching* 38:574–98.

Ladson-Billings, Gloria. 1999. Preparing Teachers for Diverse Student Populations: A Critical Race Theory Perspective. *Review of Research in Education* 24:211–48.

LeCompte, M. D., and A. G. Dworkin. 1991. *Giving Up on School: Student Dropouts and Teacher Burnouts*. Newbury Park, CA: Corwin Press.

McNeil, Linda M. 1986. *Contradictions of Control: School Structure and School Knowledge*. New York: Routledge & Kegan Paul.

Newmann, Frederick. 1992. *Student Engagement and Achievement in American Secondary Schools*. New York: Teachers College Press.

Noddings, Nel. 1992. *The Challenge to Care in Schools: An Alternative Approach to Education*. New York: Teachers College Press.

Willis, Paul. 1977. *Learning to Labor: How Working Class Kids Get Working Class Jobs*. New York: Columbia University Press.

Yerrick, Randy. 2004. Seeing IT in the Lives of Children: Strategies for Promoting Equitable Practices among Tomorrow's Science Teachers. In *Preparing Mathematics and Science Teachers for Diverse Classrooms: Promising Strategies for Transformative Pedagogy*, edited by Alberto J. Rodriguez and Richard S. Kitchen, 203–24. Hillsdale, NJ: Lawrence Erlbaum.

21

Science Learning Achievements to Leave No Community Behind

Sharon Nichols and Ruby Simon

This chapter explores issues concerning those likely to be left behind as states prepare to implement science testing in compliance with the No Child Left Behind policy. Standardized testing may encourage elementary teachers to give science instruction a more integral place in the curriculum, and resulting data from test scores might yield insights that can guide future decision making to improve science teaching in classrooms and systemically across grade levels. Test scores, however, are symptomatic as they are not causes of individual student achievement problems. Contemporary educational theories paint a more complex picture for attaining academic achievement when learning is understood as a sociocultural activity and not simply the transmission of knowledge. A sociocultural view of science learning requires teachers to look at contextual situations that challenge science teaching and learning in and beyond the classroom. Framing issues of academic achievement requires not only that student knowledge should be measured as an isolated psychological state of individuals, but also that we develop ways to address social, cultural, and political intersections of science teaching and learning. These intersections are important to consider as science testing will determine that particular ideas and ways of knowing science should be taught and evaluated based on deeper assumptions about what kinds of learners will be worthwhile products of schooling. Phrases such as "no child left behind" and "science for all" raise questions about the kinds of future citizens educational policies will produce through science education, and concerns about students whose family and community histories reflect low performance on standardized science tests.

Gloria Ladson-Billings (1994) promoted culturally relevant pedagogy as a means to support the academic success of African Americans through educational experiences centered on students' cultural and community identities, and in ways that recognize the political nature of teaching and learning. The term *pedagogy* refers to more than instructional strategies as it includes teachers' critical reflections on their philosophical views and activities—a keen awareness that their own values and biases influence practices of science education. Culturally relevant pedagogy has been shown to promote the academic achievement of diverse students.

Peter Murrell (2002) has described *African American pedagogy* as a culturally relevant framework that requires teachers to recognize teaching and learning as activities embedded in institutional, cultural, historical, and political contexts that hold particular implications for African American learners. Murrell characterizes African American pedagogy as five practices: *identity, community, participation and engagement, meaning making, and practice.* While Peter Murrell developed this framework to specifically conceptualize African American pedagogy, we see possibilities for drawing on the practices to benefit diverse learners and their communities. With this in mind, we present an approach to studying plants involving an all-black class of sixth graders as a portrait for seeing culturally relevant pedagogy in practice.

SCIENCE EDUCATION LEAVING NO COMMUNITY BEHIND

Sherry Nichols's interest to work at Vista West was prompted when she read a newspaper article about members of a black community who had worked to create flower gardens to promote pride in the neighborhood. Their activities were led by a woman named "Queen Esther"—a black woman credited with revitalizing an area previously referred to as "shanty town," where dilapidated homes had been occupied by drug users and dealers. Queen had worked with local ministers to reach out to young men to offer life and leadership support groups, and had created a local cosmetology school to provide women with vocational opportunities. The news article had captured Queen working with neighbors who were frustrated that, once again, their curbside flower gardens had been flooded by spring rains. Many streets of this area had poor drainage that readily lent to problems of flooding. Sherry saw this as an opportunity to explore science education in ways that are relevant to students' communities.

In September 2003, Sherry met Ruby Simon, a grade 6 teacher at Vista West City School—an all-black school located near Queen's 21st Street community. We quickly found that we shared deep concerns about students' low academic achievement. Vista West School's report cards, which the state requires each school to distribute, had consistently recorded "F's" in reading, writing, and mathematics since 2002. Given these trends, science testing to be implemented in 2006 would likely continue to report deficits in student learning at Vista West. The reported test results had provided little insight into reasons for students' apparent failure to learn. We felt a need to learn about the types of sociocultural issues that were challenging student learning; thus, at the end of our first year working together, we decided to use involvement in a neighborhood streetscape project as a context to practice culturally relevant science to teach about plant growth.

PRACTICING SCIENCE IN AND FOR COMMUNITY ACHIEVEMENT

Drawing on Peter Murrell's model of African American pedagogy, we gave attention to ways that student learning would extend beyond learning concepts about plant growth to also explore issues of *identity, community, participation and engagement, meaning making, and practice.* Accordingly, we started our teaching with a focus on issues of self and community identity. Ruby had students generate biographical reports about male and female African Americans whose work related to plants and/or planting in the South. Ruby shared her own autobiography of how, as a child, she longed to ride the buses to school that passed the fields where she and her family picked and chopped cotton. Class discussions about the biographies and Ms. Simon's personal story were conducted to reflect on borders that challenged the participation of these persons in science learning. *Community,* which refers to a sense of kinship, is a relationship African Americans have historically shared as a struggle for survival and freedom. These struggles persist from the legacies of slavery associated with life on the fertile dark soils of what, even today, is called the "Black Belt" region across the South. Extending from these inquiries, students wrote their own autobiographies of "My Growth as a Science Learner" to explore border crossing and identity issues they associate with learning science. Many of the students' narratives expressed having few if any memories of studying science prior to Ms. Simon's class, and some indicated feelings of vulnerability for losing friends if they were seen as eager to participate in classroom learning.

We felt a transition was needed for students to envision hope and possibilities for science education to offer personal and social opportunities. We invited two ministers and Queen Esther to visit and share about needs of the community and ways they were working to address these needs. Queen Esther and Ms. Simon shared photos of 21st Street to show the history of the area and described how community activism through practices such as the streetscape gardening had transformed the area and people. The visitors' presentations provided a context to motivate students' participation and engagement in plant growing as the following question was posed: what might make the growing of plants a *community-worthy* science learning experience? An important aspect of participation and engagement is providing supportive structures that promote students' individual and collective interests, involvement, and sustained commitment in learning activities. *Structures,* in one sense, refer to techniques for organizing classroom activities. Accordingly, Ruby continued with use of a primary organizational tool—the science notebook, which included students writing down, and a class review of, the "daily agenda" (e.g., state standards for science concepts about plants, learning performance expectations, learning tasks, and student roles) to be sure everyone understood what and how science learning would take place. We also interpreted structures as community institutions that have historically played important roles in community building, such as neighborhood black churches, grocery stores, and beauty shops. The class developed a rubric to express objectives, tasks, and evaluation criteria associated with the gardening project; however, the individualized cognitive learning targets for learning went beyond the state science objectives to determine if their experiences were culturally valuable—that is, the extent to which their science practices validated a sense of community on a personal basis and in the eyes of local citizens.

Meaning making extends beyond individual student understanding of academic concepts to involve learners in critique of symbols, signs, and other expressive modes of communication that shape their participation and engagement in educational activities in and beyond the classroom. The funding of streetscapes across east and west ends of the city revealed inequitable aspects of political practices. Queen commented that streetscapes had been established along roadways by white local commissioners to beautify their own precincts in the eastern area of the city, whereas the west side of town, an area composed of low-income black families, received no support for such enhancements; thus Queen and several members of the 21st Street community had taken it upon themselves to landscape their local curbsides.

Practice refers to students' capacity for taking action to bring about the transformation on behalf of their communities. The students expressed a high degree of interest to develop streetscape gardens, which led into science inquiry about ways to deal with the flooding dilemma that Queen and her friends had encountered. The students developed interview questions that were e-mailed to a local roadway engineer to gain his insights about the problem. He suggested that the class try to deal with soil porosity as this might pose the least expensive solution. The students carried out investigations about different soil drainage rates and explored gardening drainage solutions online. They found that a fairly inexpensive type of cement lattice could be laid down over the ground, which could contain the plants and allow water to readily drain off. Prices were compared on the Internet for the cement latticework, and using money donated by community members, Ms. Simon's class planned a budget and organized the purchase of latticework sections from a local hardware store and the delivery of topsoil to the 21st Street area. The students had grown marigold seeds in various soil types as a science investigation, with the intent that viable seedlings would be planted in the streetscape garden areas. A garden-building day was held on a Saturday afternoon. Community members worked alongside the students to place down latticework sections on several street corners. The lattice sections were filled with soil, and young marigold seedlings were transplanted into the cement containments.

In terms of evaluating learning, students conceptually learned a great deal about plant structures, plant growth in different soil mediums, and aspects of conducting science inquiry. Students were asked to share reflections in group discussion about their experience with the gardening project. One student commented,

> We fail the state tests, but those have little to do with where we live. This science project helps us to hope for more and to try to bring more of [hope] out. This sort of knowledge is real for us—how to help each other to bring about changes that improve where we live. But it also reminds us to value ourselves and where we live. That's why it's worth it, to me, to do this sort of thing.

The project enabled students to gain an understanding of science concepts, and sociocultural achievements brought benefits that extended across the student body to the local community. One of the churches conducted a Sunday celebration of the streetscape project on one of the intersections along 21st Street—attesting to community joy performed through their collective voices in song and prayers.

IDEAS TO CONSIDER

Teaching science as a culturally relevant practice is an endeavor to leave no learners or their home community behind; rather, it seeks to open up possibilities to revise science teaching and learning in ways that promote participation and meaningful science learning for all students. Achievements of science education in the streetscape project were built on the students' positive sense of belonging to a community, and on their use of agency to apply science in ways that produced collective gains. Standardized science testing of elementary students will, hopefully, encourage science to be taught across all grade levels on a regular basis; however, it must be done with a vision for how children might draw on this knowledge in ways that acknowledge and involve various social, historical, and political situations that figure into learning in and beyond the science classroom.

REFERENCES

Ladson-Billings, Gloria. 1994. *The Dreamkeepers: Successful Teachers of African American Children.* San Francisco: Jossey-Bass.
Murrell, Peter. 2002. *African-Centered Pedagogy: Developing Schools of Achievement for African American Children.* Albany: State University of New York Press.

22

Poverty and Science Education

Rowhea Elmesky

SCIENCE EDUCATION FOR ALL:
A CIVIL RIGHT

In 2001, William Tate put forth that receiving an equitable and empowering science education during the K–12 school years is a civil right for all children. The current state of science achievement as examined across racial and socioeconomic lines reveals that this right has not been actualized for children of color, especially those living in conditions of poverty. For instance, in examining the *Nation's Report Card* for science in 2000 (U.S. Department of Education 2000), the need for the acceleration of learning along racial and poverty lines is clearly pronounced. In the fourth, eighth, and twelfth grades, 66 percent of blacks, 58 percent of Hispanics, and 43 percent of Native Americans fall below the basic level of achievement in science. When studied along socioeconomic levels, the science scores across the three grade levels were also lower for students who were eligible for free and reduced lunches. These trends were true in both 1996 and 2000; in fact, in 2000, eighth-grade students of economically disadvantaged backgrounds decreased in their science performance as compared to 1996. In contrast, their more affluent peers improved their science achievement scores. Apparently, living in conditions of poverty and being of color impact student performance in the sciences, and until we acknowledge the roles poverty and race play in the ways science is taught, we will not be able to move forward toward a truly equitable state of sharing science knowledge, access, and opportunity.

In this chapter, I illuminate the ways in which being economically disadvantaged becomes compounded by race and affects the types of schools children can attend, the science instruction they receive, their placements into academic tracks, and, ultimately, their trajectories beyond high school and into society. Moreover, I highlight the attempts that have been made to provide economically disadvantaged and racially marginalized children with better access to a quality science education. Such efforts have ranged from battles for desegregation to appeals for equitable materials, facilities, and teaching forces. Most recently, science educators have articulated that underrepresented children are greatly benefited if we pay attention to classroom dynamics and the creation of classroom communities that value, respect, and capitalize on what marginalized children bring to the classroom.

POVERTY AND RACE

Urban cities throughout the United States are characterized by conditions of severe poverty on both family and neighborhood levels. As recently as 2003, the U.S. Census Bureau documented 35.9 million people as living in poverty, of which 12.9 million were children under the age of eighteen. According to the National Center for Education Statistics (2002), children living in poverty conditions tend to be located in the nation's urban centers and therefore predominantly attend large urban schools. During the 2000–2001 school year, for instance, 53 percent of the students attending the largest 100 school districts were eligible for free or reduced lunch.

Youth of color experience poverty more often than their white, non-Hispanic peers. In 2003, the poverty rate for blacks was 24 percent and for Hispanics was 23 percent, in comparison to 8 percent for whites. Most would acknowledge that there exists an essentially white superclass and a colored underclass that is continuously marginalized in schools. While poverty does affect white families, in 2001, Angela Calabrese Barton noted that economic disadvantage manifests at both the family and neighborhood levels with children of color. Isolated in communities that are predominantly segregated from whites, there is a shortage of employment options available for blacks and Hispanics. Data reveal that underrepresented groups are spatially disadvantaged in the labor market, and this is particularly more pronounced in families run by single mothers receiving public assistance. Whites continue to concentrate in suburbs, where the job availability is much higher; public transportation to these outlying areas from most of the urban centers is virtually nonexistent.

Racial segregation and unequal employment opportunities contribute to the cyclical, intergenerational nature of poverty; however, schooling and specifically science education as it has been actualized for disadvantaged populations also contribute to the social reproduction of individuals' positions in society. Science performance serves as a gatekeeper for admission into advanced placements in high school and subsequent enrollment in strong colleges and universities. Poor students of color are overrepresented in lower academic tracks, where science courses are basic, rote, and lack a laboratory component; white middle-class children are overrepresented in the highest academic tracks, where honors science courses are taught. As students of color continue to be inadequately prepared, they fail to achieve highly in science and very few have opportunities to pursue careers as scientists, mathematicians, or engineers. Science and science-related careers remain privileges for the elite, the wealthy, and the white. What, then, has been and can be done to level the playing field and allow quality science teaching to be accessible to all children?

EQUITY IN SPACE AND MATERIAL RESOURCES

Decades of legal recourse have aimed to desegregate schools so that children can receive a common and equitable education; however, truly "shared" spaces have not always arisen since neighborhoods in many large urban centers remain predominantly segregated, resulting in some neighborhood schools that are nearly 100 percent children of color. Even as recent as 2004, Kathryn Borman and her colleagues (Borman, Eitle, and Michael 2004) presented statistics confirming that current "legal" forms of racial segregation are predictors of low student achievement on standardized tests.

In addition to battles for shared space, focus has turned to ensuring the provision of equitable facilities, materials, and teacher quality so that knowledge is shared with all children. Researchers have studied schools in some of the largest districts located in urban centers such as New York City (Harlem and the Bronx), Jersey City, and Chicago. They have documented appalling facility conditions, including sewage backups and fumes, the lack of basic supplies and academic materials, and a shortage of certified and qualified teachers. In an inner-city Philadelphia neighborhood school, for example, school spaces are not conducive to learning, with classroom temperatures that are either frigid cold or sweltering hot; unsafe drinking water conditions are also an unwelcome reality. Furthermore, the teachers are given inadequate prep times and are required to teach out of field, and their input into school policy is not valued. Moreover, the science laboratory spaces are often nonfunctional in racially segregated neighborhood schools; there, the basic equipment is not present, including those related to lab safety. Fire extinguishers and goggles are missing, eyewash stations are empty, sinks are clogged, and safety showers don't work.

With the lack of appropriate and functional space, teaching science in manners promoting inquiry and problem solving is often challenging in these schools. Indeed, teachers have been found to enact what Michael Haberman (1991) referred to as "pedagogy of poverty" in urban schools. Such pedagogy refers to teaching techniques that are highly focused upon controlling the students. That is, students in science classes do more seat work and are spoon-fed information, drilled, and closely monitored. Even when computer technology is engaged, repetition tasks take precedence over inquiry and problem-solving tasks. During a 2003 study I conducted in a Philadelphia neighborhood school tenth-grade chemistry class, for example, I learned that students from the business academic track had only been allowed to read the text and answer worksheets during their ninth-grade biology class, and laboratory activities were nonexistent. Thus, when the chemistry instructor began to schedule two chemistry labs per week, students expressed excitement and interest. One male student wrote, "When it is time to go to the lab room, my interest in the class began to spark up because the labs are my favorite part of class and they are good teaching aids."

BEYOND RESOURCES DEBATES: NEEDED SHIFTS IN BELIEF SYSTEMS

Inequities in resources for science learning are not the only causes of the proliferation of poor science teaching of marginalized students. Teacher beliefs about their students impact the extent to which they search for creative alternatives to the constraints in their science teaching. When teachers see their students as capable science learners, they find ways to break through resource barriers and develop opportunities for their students to learn science. For instance, Lacie Butler, a new biology teacher in the Philadelphia inner-city school, developed labs for her students that did not require their use of advanced equipment (which was not available in the school). The conceptual content of the curriculum was not compromised by the shift of activities, nor did the teacher resort to bookwork in the face of challenge. In the same school, Cristobal Carambo, an experienced chemistry teacher, transformed an ordinary classroom located in one of the lowest academic tracks into a functional science laboratory in which the students had the opportunity to engage in dynamic and highly challenging activities such as DNA electrophoresis, growing fast plants, and comparative anatomy dissections. Cristobal gathered incredible resources through his energy, creativity, and relationships with others in education and the community at large. Both of these science teachers believed that the 98 percent black student population, with 80 percent receiving free or reduced lunches, deserved quality science instruction. Unfortunately, these cases are not the norm.

Inherent within the poorer schools serving children of color are belief systems that maintain social status. Schools, teachers, and society in general are quick (perhaps consciously or unconsciously) to place the blame for low achievement on the children or on their immediate home surroundings. Teachers express lower expectations for black students, in particular, in urban science classrooms. These beliefs are actualized when teachers routinely fail to acknowledge student ideas and ignore student requests for help repeatedly—in stark contrast to their treatment of the other students. These findings are not isolated; over and over again in the literature, black children are described as resistant, unmotivated, and oppositional in their school identities.

RECOGNIZING AND EMBRACING ECONOMICALLY DISADVANTAGED STUDENTS' RESOURCES

In part, negative notions of marginalized students' capabilities and motivation emerge and are reinforced as students and teachers interact in classrooms—especially when students engage in practices that look and sound different from what the dominant school culture defines as appropriate or respectful. However, many recent studies have found that marginalized youth are owners of a wealth of resources that can contribute to their participation in science classrooms, yet teachers—being predominantly from the dominating white, middle- to upper-middle-class population—may not understand their students' lives enough to recognize, appreciate, or make connections to their school science experiences. More often than not, miscues and misunderstandings occur as students engage with curriculum utilizing mainly unconscious dispositions or ways of being in the world.

In conducting a 2005 study, I researched "playin'" as a common practice appearing in many urban science classrooms and entailing youth's artistic use of language to respond swiftly and intelligently to another's statement. When these methods of interaction are misinterpreted and shut down by teachers, opportunities for engaging in science are simultaneously shut down, as students feel violated and disrespected by teacher responses that (to them) appear inappropriate. In contrast, in learning environments where verbal play is prevalent, fluid, and uninterrupted and occurs in conjunction with inquiry science activities, marginalized students remain focused and deeply engaged in problem-solving tasks. Likewise, Anita Abraham, a teacher researcher, and I found that the black students in her high school chemistry laboratory followed procedures both carefully and correctly while simultaneously engaging in embodied rhythmic patterns of speech and movement, including bursts of song and rap and upper bodily motion. When students are allowed to simply *be*, without constant reprimands from the instructor, they can and do monitor each other and remain engaged in science longer, thus enhancing their chances to develop positive feelings about science and their capabilities as scientists.

Children who are socioeconomically disadvantaged do not have the same opportunities to develop ways of being in the world that are accepted and privileged by the dominant class. That does not mean that the resources they hold are inferior in any sense. What it means is that deficit notions of children in poverty must be challenged. Since youth from marginalized backgrounds may lack the status and power held by the wealthy (and white), many of their practices have been overlooked, ignored, or regarded as inferior to mainstream standards. In contrast, studies have shown that students from poverty backgrounds possess dispositions that are useful in helping them build understandings of science. For instance, in my studies with marginalized youth, I have learned that, while their discourse practices can often be perceived as a confrontational challenge to teachers' authority that requires disciplinary action, these same ways of talking can manifest in student assertions, persistence, and scientific argumentation within the context of chemistry lessons. Indeed, students constantly participate in science classes in ways that connect to their experiences and realities outside of school.

While the foregoing examples represent more subtle and unconscious ways that student dispositions mix with science learning, when given the academic space, students have more consciously brought in their own forms of knowledge to understand and explain science constructs. For instance, when teaching physics to underrepresented students, I found they were able to build understandings about pitch and frequency through their shared knowledge of different rap artists who utilize pitch fluctuation for emotional expression and definition of narrative characters. Perhaps underrepresented groups will feel more comfortable and engaged with school science if more efforts focus upon connecting science curriculum to student experiences and realities outside of school.

When students are subjected to being negatively stereotyped in their science classrooms or repeatedly experience failed interactions with teachers, they may develop a history of negative attitudes associated not just with schools but also with science specifically. Thus, recognizing the capital of racially marginalized and economically disadvantaged students and embracing the ways in which those resources can help them develop school science skills are crucial for shaping a positive classroom environment. That is, when solidarity and community exist, students are more willing to engage in science activities and remain on tasks even if those tasks are not intrinsically valued. Certainly, great possibilities for expanded opportunity in science classrooms arise when we move beyond identifying "shortages" to recognizing children's skills, dispositions, beliefs, attitudes, and language resources.

TOWARD STUDENT EMPOWERMENT IN SCIENCE CLASSROOMS

Science education should not be about stratifying students or maintaining the current stratifications in place. Science classrooms should be racially shared spaces containing high-quality physical, material, and human resources. They should also be places where all children are valued and taught how to use those resources they already hold in new and empowering ways as well as taught to develop new sets of resources for becoming scientifically literate citizens.

Racially marginalized and economically disadvantaged children have long been denied access to quality science education experiences. While I recognize that there still exists great disparity in the state of science classrooms of underrepresented children, this should not prevent the emergence of science teaching that capitalizes on the multitude of resources embodied within populations of economically disadvantaged youth. However, ideological shifts must occur so that science teachers can embrace and respect the practices that underrepresented students can and do bring to learning.

REFERENCES

Barton, Angela Calabrese. 2001. Science Education in Urban Settings: Seeking New Ways of Praxis through Critical Ethnography. *Journal of Research on Science Teaching* 38:899–917.

Borman, Kathryn M., Tamela M. Eitle, and Deanna Michael. 2004. Accountability in a Postdesegregation Era: The Continuing Significance of Racial Segregation in Florida's Schools. *American Educational Research Journal* 41:605–31.

Haberman, Michael. 1991. The Pedagogy of Poverty versus Good Teaching. *Phi Delta Kappan* 73:290–94.

National Center for Education Statistics. 2002. *Characteristics of the 100 Largest Public Elementary and Secondary School Districts in the United States: 2000–2001.* http://nces.ed.gov/pubs2002/100_largest/highlights.asp.

Tate, William. 2001. Science Education as a Civil Right: Urban Schools and Opportunity-to-Learn Considerations. *Journal of Research in Science Teaching* 38:1015–28.

U.S. Census Bureau. 2003. *Poverty: 2003 Highlights.* www.census.gov/hhes/www/poverty/poverty03/pov03hi.html.

U.S. Department of Education, National Center for Education Statistics. 2000. *The Nation's Report Card: Science 2000.* Washington, DC: Author.

23

Gender Equity Issues

Kathryn Scantlebury

Gender is a social construction and can influence students' science education in multiple ways. A person's gender is constructed in relation to other social categories such as ethnicity, class, race, religion, and language. For example, for students from underrepresented groups, participation in science requires a larger cultural shift between personal (home) and public (school) spheres than it does for students from the white, middle-class, mainstream culture. The shift from home to public spheres may be easier for African American females, who are less constrained by traditional gender roles and community compared with girls from other underrepresented groups, such as Latinas.

SCHOOLS AND TEACHERS

A student's school and teachers can influence her or his engagement with science. Schools, and other education institutions, rely on the interactions between individuals and groups to function. Gender becomes a factor in shaping those interactions and relations. Schools often reinforce rather than challenge or change gender stereotypes. For example, twenty years after Title IX of the Elementary and Secondary Education Act legislation that banned sex discrimination in education programs and activities, researchers noted that colleges and universities had a "chilly climate" for women and that public schools "shortchanged" girls. According to a 1998 report from the American Association of University Women, girls' schooling experiences vary depending upon their socioeconomic status, geographic location, ethnicity, and/or disability. More recently, comments by Harvard University's president that innate sex differences may contribute to fewer female faculty in the sciences resulted in national and international discussions on how cultural factors are more likely to explain women's participation in science rather than biological differences between females and males.

Gender issues in science have addressed pedagogical practices, curriculum choices, assessment techniques, and students' participation patterns in science courses and careers at all stages of the formal education pipeline and in informal settings. Since gaining equal access to science education, girls and women comprise at least 50 percent of the majors in many science programs; however, inequities between females and males continue to exist.

PEDAGOGICAL PRACTICES

Target students are students who demand the teachers' time and attention. At all levels of schooling, target students are typically white males. Teachers are more likely to ask boys questions, particularly when those questions are complex, abstract, or open-ended. Boys, more than girls, will volunteer answers to teachers' questions, call out answers, and seek teachers' attention and time through questions or noncompliant behavior. Teachers reward girls when they are helpful, compliant, and quiet. When girls do not conform to this gender stereotype of femininity, they are labeled "troublemakers." For example, African American girls are encouraged by their families to develop self-determination and assertiveness, these traits are not encouraged in schools, and the girls experience a disconnect between the culture of home and that of school.

Another outcome of gender stereotyping is the expectation that girls will assume nurturing roles toward other students. Teachers ask girls to take on mothering or caretaking roles to students, often males, who have fallen behind with learning because of inattentiveness, absenteeism through truancy, or in-school disciplinary procedures. Rather than extending and expanding their personal knowledge of science, girls are asked to teach their male peers, that is, to assume "other mothering" roles.

Participation in inquiry is an important activity that can assist students to learn science. This can involve students in using equipment and participating in verbal interaction. Besides dominating the teacher, target students may control other resources in the classroom, such as laboratory equipment. Boys appropriate science equipment, and, without intercession from teachers, they may not share the equipment with their peers. In laboratory groups, girls can be relegated to roles such as data recording or reading instructions. A teacher's conscious intervention can minimize this unequal sharing of resources.

Frequently, girls and boys have different educative experiences in science classrooms. Group work can engage more students in science, but teachers must monitor the interactions between students in those groups to ensure that all students are participating and that one student is not dominating the group. Most students prefer that science is taught using groups and hands-on activities. Students, especially girls, dislike lectures, worksheets, and "busy" work assignments.

Girls often excel in single-sex education settings. Research conducted in public schools in other Western countries such as Britain and Australia have reported the positive impact on girls when they learn science in single-sex settings. However, administrators dismantled these arrangements because teachers reported that "boys" in single-sex settings are unteachable and require girls' civilizing influence.

The acceptance and consistency of traditional gender roles, boys being rewarded for assertive behaviors, girls for compliance and nurturing practices, in schools are often invisible to students and teachers, and the subtle inequities in classrooms are barely noticed by the participants in classroom life. Even when these practices become visible, students, teachers, administrators, and parents regard them as normal, and these equity-related subtleties continue to contribute to gender differences in science.

CURRICULUM CHOICES

Educators are concerned with students' declining interest in science, and in particular, that academically able girls and women show an ongoing pattern to seek careers in other areas that are not related to science. While girls' school science achievement has increased during the past decade, their overall attitudes toward science have remained negative. In general, girls prefer other school subjects and perceive that they are better in those subjects compared with science. Girls prefer studying subjects that they perceive as having value and relevance in their lives. Often, science is taught without an emphasis on how the subject connects to the "real" world. If girls are interested in science, they prefer the biological sciences rather than the physical sciences. This pattern shows that students' preferences in science are aligned with their gender identity, that is, biological science is perceived as more feminine than the physical sciences. For example, the percentage of girls taking Advanced Placement (AP) classes in science has increased significantly over the last decades; the one exception to this pattern is high school physics, where more boys than girls are enrolled.

POLICIES

The No Child Left Behind Act has tied state funding for education to student performance on high-stakes tests. States are mandated to provide annual reports of academic achievement by almost every social category *but* gender. Moreover, data are rarely disaggregated by gender, social class, and race. Various school and community factors may influence student achievement, such as teachers' certification, content background, pedagogical practices, and students' home and community support. These factors have different impact on the achievement of female and male students. For example, a supportive home environment has a positive impact on the science achievement of urban, female and male African American students. While peer support can influence African American girls' science achievement, gender, social class, and race can impact students' science trajectories, career decisions, and participation; and without data to document how these social categories interplay and interact, our understanding of gender and science is limited. It is possible that without continued monitoring of gender differences, the gains made in this area could disappear.

PARENTS

Another aspect of how a student's gender can influence her or his involvement in science is parental perceptions and expectations. Middle-class, white parents perceive that science is more difficult and less interesting for their daughters. One outcome of this perception is that fathers use more cognitively demanding speech with their sons when talking about science. Parents' attitudes toward their children's science interests and abilities strongly influence the students' interest and self-confidence. Parents can promote their children's interest in science by buying toys that develop science skills; providing children opportunities for science-related experiences such as visits to museums, aquariums, and zoos; and encouraging children to pursue science in school.

OVERCOMING GENDER DIFFERENCES

The stereotype that science is a masculine and white endeavor still exists, and a learner's gender can influence her or his educative experiences, attitudes, interests, and engagement in science. For these disparities to disappear, we must continue to challenge the assumptions of "who can do science" from parents, teachers, other educators, and policy makers. Teachers can proactively challenge the stereotype through their pedagogical practices. For example, teachers can reflect on their practices through simple questions, such as the following: which students are engaged with the curriculum? How does the teacher use questioning techniques to engage students in science? Does the teacher ask girls and boys complicated questions? Are target students evident in the class, and if so, how does the teacher interact with those students? Does the teacher use a variety of pedagogical and assessment practices?

By being cognizant of science's strong masculine stereotype, parents can encourage their daughters to engage in science by encouraging their participation in out-of-school science and/or tinkering experiences. For example, those experiences may include encouraging girls to engage in activities such as visiting museums, attending science camps, and dismantling electrical appliances or machines to explore how those objects work. These activities develop science dispositions, such as manipulation of equipment and spatial-visualization ability, that students can access and use in school classes.

The differences within gender are greater than the differences between genders. Gender can influence students' engagement with science. How social categories such as ethnicity, class, race, religion, and language interact with gender and how those interactions impact students in science comprise an ongoing research area.

REFERENCE

American Association of University Women Educational Foundation. 1998. *Gender Gaps: Where Schools Still Fail Our Children*. Washington, DC: Author.

24

Understanding Agency in Science Education

Angela Calabrese Barton and Purvi Vora

INTRODUCTION: KOBE'S STORY

Seven years ago, I got to know Kobe, a 16-year-old African American man, through an after-school program where youth learned science through converting an abandoned city lot into an urban garden.

Kobe had been living with his mother and younger siblings in a long-term homeless shelter in New York City. Kobe's story is interesting because depending upon which aspect of Kobe's life you examine, you see a different person. If you viewed Kobe through standard school terms, you saw a young man who was failing science (and other subjects), who was more interested in sports and gangs than academics, and who eventually dropped out of school. While Kobe expressed a clear desire to one day become "rich" and make it into the NBA or NFL, he could not participate on school sports teams due to his low grades. If you viewed Kobe through the lens by which we got to know him, you met a young man struggling to help his family survive, who was critical of outside efforts to "help" his community in ways that did not attend to the reality of his community, and who engaged in science when it valued his identity and context. Kobe served as the primary caregiver for his younger siblings when his mother was away, and often missed school due to these responsibilities. Kobe valued his affiliation with his gang because it offered financial support and a place to belong, neither of which he felt he got at school.

When the lot transformation project first began, Kobe resisted involvement in the project. He argued with his more involved peers that their efforts were wasted because any changes they made to the lot could be negated by gang activity and vandalism. Kobe was particularly concerned that even if they could prevent vandalization, they would not have the expertise they needed to make the lot a safe playground for younger children.

But Kobe ultimately did become involved in the lot project, and his involvement was profound. Kobe first became involved through participating in the video documentation of the lot project. He was interested in technology, wanted access to sophisticated equipment, and found the flexibility of the documentation project more suitable for his unpredictable schedule. He later became involved in "community days" (where the community was invited to work on the project on Saturdays), which offered him support by providing free child care and lunch. Kobe, through these peripheral activities, became part of the group of teens working on the project, and eventually played a more central role in all of the project activities.

Participation in the lot project represented a big shift for Kobe. While participation was grounded in his desire to gain access to technology, child care, food, and fun, it was also rooted in a desire to see real change happen in a neighborhood neglected by the city. Yet to be successful meant that Kobe had to move beyond these desires to learn new science ideas and skills. Each new step of the project required that youth learn something new in science. Kobe, too, had to learn how to identify the investigatory questions and concepts that allowed them to transform their critique into a real plan of action. He needed to learn how to design and conduct scientific investigations, such as documentation of the state of the lot and determination of appropriate cleanup activities if a garden was to thrive. He needed to study technology and mathematics to improve the investigations (e.g., issues of scale, measurement, and modeling). He needed to formulate and revise scientific explanations and learn how to weigh evidence in order to make sound decisions. Finally, he had to learn how to effectively communicate the group's concerns and findings to a larger community to acquire financial support and greater community participation.

Kobe's participation transformed his beliefs about what he could do with his life: he began to believe that he could have science as a backup career if the NBA did not work out. He elected to go back to school to learn more science and get a high school diploma to make this career option viable. This is a profound shift, for it suggests that Kobe's vision of who he was and what he could accomplish was expanded through this experience, in a way that empowered him enough to return to school—a place he found distancing and unsupportive.

LEARNING FROM KOBE ABOUT THE IMPORTANCE OF AGENCY

As Kobe's story suggests, we are interested in how youth work to bring about change in their lives—or, in other words, how youth craft a sense of agency. Over the past decade, we have been working with urban youth to better understand how they come to engage in science both in and out of school settings. While the primary focus of our attention has been on "how and why" youth engage in science, we have noticed over time that a big part of youth's purposeful engagement is related to their agency.

But, what is agency? Many of us—teachers, researchers, parents, and students—often talk about how it is important for students to feel empowered in school and in science. We also talk about the importance of students having opportunities to take control of their lives and their science learning.

We all might agree that agency is important, but we do not always have a clear understanding of how students learn to develop a sense of agency or to expand what agency they already have. Furthermore, we may also lack the ability to see or understand when students are trying to express agency, muddling their attempts to gain more control over their science lives.

In a general sense, human agency can best be described as individuals or groups acting upon, modifying, and/or giving significance to the world in purposeful ways, with the aim of creating, impacting, and/or transforming themselves and/or the conditions of their lives. And, in order to do so, individuals creatively adapt resources available to them to attain new ways of acting and being within whatever context they confront.

But what does this view of agency really mean? How do youth use the contexts of science learning to act upon or give significance to their worlds in ways that allow them to transform those worlds? And when we talk about agency in science education, we also wonder, *how does science matter?*

Kobe's story shows us that the lot project was initiated in response to the youths' critique of their neighborhood and the city's lack of attention or care for their neighborhood. The project was successful because the youth strategically and creatively drew upon a wide range of resources to engage the community in doing science that made their community a safer, more beautiful place to play. For example, Kobe and his peers created a series of venues in which the lot improvement project took place: regular after-school meetings focused on science investigations; weekend community days focused on cleanup, planting, and community involvement; a video documentary recorded the development of the project and taught other youth how to conduct similar projects; and a binder recorded all actions, investigations, and data collected. Kobe and his peers were clearly stakeholders in the process, acting as both agents and subjects in the process.

Yet, the case of Kobe can be limiting because it is grand in scale and takes place in an out-of-school context. One of the questions we asked ourselves was in what ways do students develop a sense of agency within the context of school science, perhaps on topics much more everyday than changing the neighborhood?

QUACEY AND THE SCIENCE FAIR PROJECT

Quacey was a sixth-grade student in a high-poverty middle school in Harlem focused on math and science. His school was newly opened in the district's efforts to replace a school that had been on the failing list for four years. Quacey was liked by his teachers and his peers, though students sometimes made fun of his "big lips."

Quacey deeply respected his science teacher. He believed she was smart and caring. He noticed that she worked hard to make her classes enjoyable, to listen to and respect students, and to help her students learn as much science as possible. Like many children his age, Quacey believed his science teacher was "never wrong." When prompted, he could not come up with a single time when his teacher did not know an answer to a student's science question. He also believed that he should never contradict his teacher, especially when it came to science ideas. This was part of respecting her but also part of his belief that she "knew science." He worked hard in her class to meet her expectations and to please her, which is not always easy given the peer pressure.

Quacey also loved to build. His father and his brother are construction workers. In his free time, Quacey plays with screwdrivers and motors, and builds cars. Quacey was also articulate about the science behind the things he built, drawing upon his experiential knowledge to explain how things work: "Well the [car] motor, uh the motor got like magnet inside of it... It like a magnet.... [I]t got metal around it and it got magnet and it got like this thing inside."

As part of district requirements, Quacey's science teacher needed to have her students complete a science demonstration project that would meet city standards for inquiry science. These demonstration projects are similar to "science fair" projects in that students complete a project that demonstrates the scientific method and present their findings on a poster board. The norm in this district was for students to write up their findings to a lab designed as part of regular classroom learning. So while students conducted independent investigations, they had little autonomy in the design of the projects. Further, while students in the district were required to complete a demonstration project, they were not required to exhibit their project at the school or district level.

Quacey's teacher, Ms. R., decided that in collaboration with the school's other two science teachers, the students would prepare science fair projects to meet the needs of the District Exposition Project and share their work with the school and neighborhood community. Further, Ms. R. decided that the crux of the science fair project would be an experiment: each student would conduct his or her own investigations and present them in the standard demonstration format for the school to view. She wanted her students to have authentic projects, but she also wanted them to have structure, since they had not participated in such an event before in that school.

Ms. R. offered three organizing structures to support students in developing their projects:

• A list of experiments the students could select from, based on her own research into experiments that were "inexpensive" and "simple" and that could be measured, rated, or quantified

- A guideline for preparing the research paper and poster
- A series of worksheets aimed at helping students through each of the primary steps in their investigation (i.e., developing a hypothesis, designing the experiment, etc.)

After an initial week's work of intensive instruction/activity on the projects, she provided one class period a week for several weeks to allow groups to work independently on their projects while also having access to school resources.

During the first week of in-class science fair work, Ms. R. distributed the list of experiments to choose from and offered the groups time to investigate the list and settle on an idea. She also offered them a worksheet to guide them through how they would transform the research topic into an experiment. Despite his interest in the science fair, Quacey was dissatisfied by the list. He was keenly interested in space exploration and wanted to build something. It was therefore particularly interesting to us that Quacey, who loved his teacher and felt she was always right, rejected the list and proposed to build a space rocket, an idea that with some refinement was ultimately approved by his teacher.

How did Quacey get to pick something off of the approved list, and why does that matter? When we take a close look at what happened over the course of the science fair project, we notice that Quacey strategically adapted the resources he had available to him to convince his teacher to allow him to do something different.

What we think is particularly interesting is not just "what" resources Quacey used, but also "how" his group strategically integrated those resources into what he already knew or cared about. In particular, we noted two important strategies: using the tasks to redefine their space for work, and negotiating and prioritizing resources available to them. We discuss both points together since they are interrelated.

Quacey first sought approval of their topic by adapting the task to fit his interests and by using the expectations of the figured world of his classroom to support their ideas and interests. For example, in presenting the topic of building a Mars Rover Rocket Ship for his project, in both written and oral formats, he strategically drew upon school-sanctioned resources to do so. He demonstrated a connection to a topic previously studied in class in a way that validated his topic and showed it as worthwhile and scientific. In his discussions with his teacher, he referenced the class' previous study of the Mars Rover, a topic his class briefly covered months earlier during a NASA rover rendezvous with Mars. He also presented written evidence for scientific sources of information he could draw upon (i.e., NASA website and science books), and showed knowledge of key science ideas involved (e.g., pressure and liftoff, H_2 as fuel, and electricity). He also framed the building project as a design experiment, by offering a hypothesis: "*My hypothesis is my experiment is that our rocket ship will vibrate and the light will be flashing and the perpeler* [propeller] *will spin.*"

While the teacher conditionally accepted Quacey's idea, she was firm in her resolve that his project needed to be "more experimental"—that he should offer evidence that his rocket was a success or failure. As a result, Quacey and his team continued to negotiate aspects of their project to maintain the support of their teacher (i.e., how to frame the project as an experiment) while holding to those dimensions of the project most important to them (i.e., building something). Quacey adapted two key teacher suggestions: he referred to his rocket as a "model," and he modified the protocol to include multiple tests in order to offer quantifiable data. He and his team also worked on designs for two different models in order to compare data.

As a result of these strategic negotiations, Quacey was able to maintain building as an emphasis of his project but at the same time learned to view building as a process of design and experiment. He was able to keep as central both his identities—that of a builder and of a good science student. Through the activity of building, he was able to validate his efforts to draw upon nonschool-based resources for completing the project—parental expertise on building and on using the Internet, parental views on aspects of science that are important, personal experiences building, and social alliances for drawing in various forms of individual expertise (i.e., the group expanded to four students, and one person was the artist to create the final model sketches, Quacey was the lead builder, while two others were the researchers and writers).

CONCLUSIONS: WHAT DO QUACEY AND KOBE TELL US ABOUT AGENCY?

Both stories are good examples of agency because they show us how youth draw upon resources they have access to in order to bring about change—whether that change is large, like transforming a lot into a community garden, or small, such as changing one's topic for a science fair project. Both stories also shed light on four important qualities or dimensions of agency.

First, they show us how agency is not something that can be ascribed to an individual but rather it is *an expression and a process that changes from moment to moment*. In other words, neither Kobe nor Quacey possessed or dispossessed agency. For example, Kobe began to learn how to express his agency in the lot project after learning more about how a lot can be transformed and how he could access important resources to do so. But while he gained a deeper sense of agency in the lot project, this agency did not translate well into his school science setting, where he continued to struggle to succeed.

Second, the two stories show us how agency is grounded in *intentioned activity*. In other words, how the students strategically drew upon resources was grounded in a critical awareness of their situation (i.e., a need for a safe place to play, and a desire to having building respected in science class), a desire to assert their intentions toward acting upon that awareness, and a need to disrupt the expectations and stereotypes that dominate their experiences.

Third, in enacting intentioned activities, students actively *position them-selves* as authors, as capable knowers and doers of science, and as individuals entitled to positions of power. As authors of their contexts, students creatively use resources to disrupt normative patterns. Quacey worked to keep his identity of builder alongside that of a good science student. Likewise, Kobe, who was a gang member and failing at school, became a leader on the lot transformation project. As capable knowers and doers of science, they reenact their experience to legitimize their knowledge and position. Kobe's experiences as primary caregiver played into developing important criteria for the lot development. Likewise, Quacey's home knowledge about building became central to the design of his project. As individuals with power, they become both agents and subjects of change: in other words, they become primary stakeholders rather than passive participants giving significance to that activity.

Fourth, the stories show us that science plays an important role in serving as both a tool and a context for bringing about change. Both Kobe and Quacey used science as a tool for change to investigate issues of personal and social importance, to explore issues of equity and fairness, to support critiques, and to prove and argue their points of view. As a context, science not only became the knowledge community where their experiences were legitimized but also itself was allowed to be changed by those experiences. In other words, "what science was" shifted for Kobe and Quacey as their experiences took on new meaning within a science context.

Thus, agency in science education is an important part of becoming scientifically literate. What students "know and can do" with science in their lives is only as important as their ability to use that knowledge to transform the conditions of their lives.

25

Building Rovers to Bridge the Achievement Gap: Space Exploration at School and at Home

Alberto J. Rodriguez

OUR VERY FIRST ROVER

It's been over two hours, and we're still building our rover. I've never seen my daughter and my son so intensely involved in making something work. Our rover has arms to grab things, and a small motor that looks like a tiny air conditioner unit operates them. Two more motors give power to our back wheels. A capsule with a toy astronaut is mounted on the base of our rover, and although it moves slowly, my kids smile widely when they see that the product of their labor has life—it makes noise, it moves, and it grabs things. They of course start taking turns to see if they can maneuver the rover's arms to pinch each other's fingers. This is not exactly a top NASA mission, but it was certainly making them laugh and enjoy their creation. I knew that this toy was going to provide us with many more hours of fun together, and at the same time allow my children's imagination and creativity to take flight.

That was over twelve years ago, when I bought my children their first Robotix kit, which is a robot-building kit that contains many connecting parts and joints, a set of wheels, motors, cables, and a battery-operated control unit (Robotics and Things n.d.; see www.roboticsandthings.com).[1] Now that my son is about to graduate with a degree in chemistry from Vanderbilt University, and my daughter is finishing her first year of the biology program at San Diego State University, I find myself building robots again. However, this time, I am building them with schoolchildren from grades 4–6 as part of our research projects. As a science education university professor, I am committed to investigating how education can be made more student-centered, more equitable, and more culturally and socially relevant to all children.

Therefore, in this chapter, I begin with a discussion on the importance of making science more equitable and accessible to all children. This is followed by a detailed description of the student-centered, problem-solving scenario we have developed to make the learning of robotics and school science more socially and culturally relevant, and more engaging for all students.

THE NEXT GENERATIONS OF SCIENTISTS, ENGINEERS, AND INNOVATORS: WHO WILL THEY BE, AND WHAT ARE YOU DOING TO SUPPORT THEM NOW?

We all need to be concerned about the pervasive gap in student achievement and participation in science and mathematics between Anglo-European students and students from diverse cultural backgrounds (e.g., Latinos/Latinas, African Americans, and Native Americans). Despite more than fifty years of school reform efforts and research since the *Brown v. Board of Education* (1954) desegregation decision by the U.S. Supreme Court, this gap in achievement and participation continues, and it can also be observed along socioeconomic and gender lines (Rodriguez 2004). One factor that contributes to these inequities in educational opportunities is access to instructional technologies. Many researchers have clearly indicated that where there is a gap in student achievement, a digital chasm also exists between the same students in terms of access to learning technologies (CEO Forum on Education and Technology 2001).

There is hope, however, as recent studies have shown, that the academic performance and participation of students from diverse backgrounds and/or from low socioeconomic status improve significantly when at least one of the following approaches is implemented in the classroom: teachers having high expectations of *all* students; students conducting hands-on, minds-on, and student-centered activities frequently; and students having access to the effective use of learning technologies.

Researchers Dick Corbett, Bruce Wilson, and Belinda Williams (2005) found that students from culturally diverse and economically disadvantaged urban schools improved their academic work when their teachers held them accountable to high expectations. In interviews with the students, these researchers also found that the students themselves appreciated their teachers' efforts and saw them as "caring" for them. In another study, Fred Newman and his associates (Newman, Bryk, and Nagaoka 2001) found that students whose teachers gave them intellectually challenging assignments showed a gain on the Iowa Test of Basic Skills that was 20 percent greater than the national average. Newman and his associates conducted the study in over 400 classrooms in Chicago and analyzed over 2,000 assignments in writing and mathematics.

Related to intellectually challenging assignments that require application of knowledge and not just rote memorization is the need to expose students to more authentic opportunities to do hands-on and minds-on science. For example, Wenglinsky (2001) suggests that "students whose teachers conduct hands-on learning activities outperform their peers by more than 70% of a grade level in math and 40% of a grade level in science" (8). A research brief prepared by the American Educational Research Association (2004) also supports all the research findings mentioned above and adds, "[Student] performance improves when all students have the opportunity to learn the same challenging curriculum, marked by high standards and expectations" (4).

In another research brief produced by Apple Computers (2002), the findings of various studies are summarized regarding the effective use of learning technologies. These studies suggest that when teachers use learning technologies often and effectively in the classroom, students—including those from diverse and economically disadvantaged backgrounds—are more motivated and engaged. As a result, they spend more time learning in the classroom and applying what they learn in and outside school.

The rover activity described below combines all of the suggestions for increasing student achievement and participation mentioned thus far, and it has provided us with multiple opportunities to assess firsthand how students from culturally diverse backgrounds react to these approaches to teaching and learning in our own research projects (for more information, see findings from the Maxima Project at [Maxima 2004] and findings from the I2TechSciE Project [Rodriguez, Zozakiewicz, and Yerrick 2005]).

BUILDING YOUR OWN PLANET EXPLORATION
ROVER AT HOME OR AT SCHOOL

In this section, I will first provide a version of the problem-solving scenario we have used in our work in urban and culturally diverse schools. This scenario provides a "hook" for the students to get more engaged with the rover activity, apply what they know, work collaboratively, and make socially relevant connections between the science content and how it can be applied in real (or potentially real) situations. The same scenario could, of course, be used at home to start your own space exploration adventure. The scenario is followed by some suggestions on how to build a rover. After building so many versions of the rover with elementary, middle, and high school students, our designs have continued to get better and more sophisticated, so dedication, imagination, and a sense of adventure are the key ingredients to many hours of collaborative learning. Finally, some specific suggestions are offered for making the implementation of the activity more purposely culturally inclusive, gender inclusive, and socially relevant.

Problem-Solving Scenario: Lost in Space

Note that before presenting students with this activity, a moon or planet scenario must be built. We have found that red butcher paper works well for creating the landscape (including mountains, hills, and valleys). Volcanic rock (easily found in garden stores) works well to create the illusion of scattered rocks and boulders. A curtain made out of the same paper as the scenario or some other appropriate colored material should be used to keep the scenario hidden from sight. The scenario does not need to be very big, if you have limited space. Our latest scenario was 5 feet by 10 feet, and the students found it challenging because students were allowed to see the scenario only through the "eyes" of the rover (the wireless camera). The activity could be more student-centered by having two different classes create a scenario for each other. The key here is to keep the identity of the moon or planet represented by the scenario a secret. The goal is to encourage students to use the features of the planet or moon, and to identify it correctly.

Teacher reads the following to students:

Oh, no! Your spaceship is out of control!!! Soon after taking off, a huge solar flare hit the Earth, disrupting all types of communication. Your ship is spinning into outer space at a high speed, and you can't tell what's up or down anymore. Smoke is coming out of your control panels as flashes of light blind you.

You and your team members wake up after being unconscious for many hours, and you are not sure where you are anymore. Your sensors are fried, and you can't contact NASA for help. One of your team members looks through the cockpit window and points out that your ship is caught in the gravitational pull of a solar object. You can't tell if it is a planet, a moon, or some other solar object far from earth. It might be Venus, or it might be Mars or something else. What are you going to do? How are you going to contact NASA for help when all communications are down?

Your team decides to start making repairs to the spaceship controls and radio right away, but you know that this is going to take some time, and you need to contact NASA. Therefore, you decide to send your remote-controlled rover to explore the surface of the solar object in hopes that you could identify it. You also know that previous missions have left communication antennas and other supplies in storage on nearby planets and moons. If you are lucky, this solar object may just have what you desperately need to repair your ship and call home. You just have to figure out where the best area to land your rover should be. If this were a planet or moon that was previously explored by NASA, what would be the best place to land?

Once the rover gets to the solar object, you will be able to maneuver it using the remote controls from your ship. It was lucky that your rover and its controls were built separately from the ship's main control; otherwise, you would really be stuck. Once you land the rover, you will be able to see what the rover "sees" thanks to the video camera on the rover. You really need to be on your toes and stay sharp. You don't know what you will find or how long your rover will last in that environment. . . .

Directions for Teams: Use what you learned about the planets and their moons to identify the solar object your rover is exploring. Write down as many key features about the solar object as you can. For example, are there any mountains, craters, active volcanoes, ice caps, and so on? What is the color of the dirt, and what is the temperature? Discuss your opinion with members of your team, and then write the name of the solar object and your reasons for this choice. If there is disagreement among your team members, that's OK. Write both responses and why you disagree with your team members' choice.

Building Your Own Rover

For our rovers, we prefer to use the Robotix kits because we find the building components to be much easier to handle than those produced by other companies (e.g., Lego). The plastic building components of the Robotix kits are creatively designed, and the hundreds of pieces in every kit come with connecting joints, a set of wheels, motors (with high and low speed), cables, and control units. Some kits come with launch sounds, lights, and remote controls. I recommend buying the kits with remote control because, even though they are more expensive, they increase the range of operation of your rover.

We have found that a rover moves too slowly when using the motors and wheels that come with the kit. This is fine if you are building a rover to use at home, but if you want to use one with students in your classroom, the rover may move too slowly for most students. Therefore, I suggest buying a midsized radio-controlled car, and then taking off the top (usually, unscrewing four screws is all that it takes). In this way, you are left with an excellent base onto which a rover can be built. By using a glue gun and some of the flat building pieces from the Robotix kit, we have been able to build a versatile base for any kind of multipurpose rover our imagination has conjured up. Note that you only need to glue the flat pieces onto the remote car to create the base of your rover. It is better to connect all other components using the normal connecting joints so that you can build, expand, and take apart your creation as often as needed. It is also important to make sure that the frequency of the Robotix remote-control unit is different than that of the remote-control car.

Depending on your budget and/or purpose, rovers could be built with or without probes to gather scientific data from its surroundings. Figure 25.1, for instance, shows the Aracas IV, the latest rover we have built. As can be observed from left to right, the rover has a grab-arm (included in the kit); a magnetometer produced by Pasco (1996–2005) in the center; a wireless, remote-controlled, mini–color camera (X10 1997–2006); and a temperature probe produced by Vernier (n.d.). The magnetometer probe is connected directly to an Apple MacIntosh laptop via a long USB cable. The temperature probe is connected first to the DataPro. This is the Vernier data-collecting unit that comes with the probes (the DataPro is mounted on top of the rover with velcro). This unit can have up to four different probes connected to it at a time,

FIGURE 25.1   The Aracas IV: A moon/planet exploration rover.

and it allows the user to send simultaneous data to the laptop for analysis via only one wire. For our problem-solving scenario, we were only measuring changes in the magnetic field and temperature. To this end, we hid strong magnets and created hot spots in various areas of the planet scenario we built so that when the students were maneuvering the rover near these areas, they would notice data spikes on the computer screen. Needless to say, this created a lot of excitement as students began to use their science content knowledge to speculate reasons for the observed phenomena. Thanks to the wireless color camera, the students were also able to maneuver carefully around "dangerous" hot spots and unknown terrain. Throughout the activity, the students were required to use their scientific knowledge to explain the observed phenomena and describe the geological features of their unknown solar object.

One of the problems we found with using the bottom part of a remote-controlled car to build the rover is that now the rover could move very fast. In order to encourage students to get used to driving the rover at a reasonable speed for exploration, we created "dangerous zones." In other words, if the students drove the rover too fast, certain areas of the scenario had rock slide zones (boulders made out of the same butcher paper used to construct our scenario). The basic rule was that if the rover "crashes," that group will lose their turn (the mission failed), and another team will have a chance to explore.

We found that having the scenario built in a separate room, and then having two to three teams at a time in the scenario room, provided for deeper and more meaningful engagement. While one team was engaged in the mission, another team was observing and planning, and the third team was practicing how to control the rover with another replica we had available. It is important to note that the scenario was covered by a curtain, and although the scenario was small (5 feet by 10 feet), the students found the activity challenging because they could only see what the rover "saw" via the remote camera. Since this image was transmitted to a TV monitor, and the students were controlling the rover remotely, they had an opportunity to appreciate the difficulties associated with space exploration and the need for intelligent and practical technological design. We were quite impressed with the variety of engineering suggestions proposed by the students for improving the technology and design of the rover so that there could be more control and maneuverability, and more opportunities to gather and analyze data. We were also fascinated by the students' creativity to explain the observed phenomena and by the arguments they used to defend their position on whether the planet was Mars, Venus, or some other solar object. One student, for example, argued that the reason we were observing localized spikes in the magnetic field might be due to residues left by meteorite impact on the area.

Other Strategies for Making the Activity Culturally and Socially Relevant

Any teacher who attempts to conduct the problem-solving scenario in collaborative groups as described above is already making her or his class more culturally inclusive and socially relevant. On our research projects' websites, we describe several strategies for making science more purposely gender, linguistically, and culturally inclusive (for more information, see Maxima 2004; Rodriguez, Zozakiewicz, and Yerrick 2005). Here, I briefly mention some of the approaches we found to be most effective thus far.

Learning Centers

One way to maximize limited resources and equipment and to maximize learning is to create learning centers. At each center, students may have specific tasks or problem-solving scenarios like the one mentioned above. For example, one of our learning centers focuses on the contribution of women to space exploration. At this center, students are required to read documents that describe the history of women in space, and then engage in a discussion with their teachers and peers. Students are asked to reflect on and discuss questions, such as why it took so long for a woman from the United States to join a space exploration mission (1983) when the former Soviet Union sent their first female astronaut twenty years earlier. Another question is whether they have heard about the first Latina and/or first African American U.S. astronaut, and if not, why not? Students often make very insightful comments and point out that their books do not have pictures of these female astronauts or other scientists that look like them, and they wonder why. Students are encouraged to discuss what they learned in class with their parents and to continue reflecting about gender-related issues in today's society, and the role they could play in making society more equitable for everyone.[2]

Same-Gender Grouping

We have found that in the upper elementary grades (especially in grade 6), girls prefer to work in same-gender groups. They find working with other girls more productive and more focused. Interestingly, our research has shown that boys often prefer to work in mixed groups because girls help them "get the work done" and "fool around less" (Rodriguez and Zozakiewicz 2005). This points to the need of not taking for granted how students are assigned to work in groups. Teachers must monitor closely the students' productivity and level of collaboration within groups in order to ensure that traditional gender dynamics are not getting in the way of learning. What this involves in paying attention to all learners, but especially looking to see whether or not there are trends that suggest that girls are not as involved or are not achieving in the same ways as their male counterparts. Of course, other factors, such as race and language resources also are potential borders that can produce inequities.

Using Technology Involves More than Just Using Computers

Unfortunately, instructional technology is usually perceived as only involving the use of computers in the classroom. Even worse is the perception that computers are only good for doing searches on the Internet or typing reports. It is important to help students realize that other forms of technology (e.g., digital video cameras, educational software and games, web-based educational programs, and data collection probes) could be used in combination with computers to add new dimensions for more meaningful learning. For example, a trip to the zoo could become a more meaningful and socially relevant trip by having students take pictures of not just the animals but also how their habitats are constructed, how the public is protected from the animals and vice versa, how classmates react to the animals, and so on. Students could then download the photos into the computer and create either photo stories or reports that could motivate and facilitate learning for everyone (including second language learners).

MOTIVATING ALL CHILDREN TO LEARN USING ROBOTICS

By not taking for granted how students are assigned to work in collaborative groups, by providing multiple opportunities for hands-on and minds-on learning, by having high academic expectations of all students, and by providing students with frequent and meaningful access to learning technologies, we could make science education more equitable, culturally, and socially relevant, and, indeed, more enjoyable to all students. The pervasive gaps in student achievement and participation for the last fifty years clearly demonstrate that it is going to take a lot more than good intentions and superficial policies to effect long-lasting change. Perhaps one way to start making education more equitable at school (or even at home) is by conducting problem-solving scenarios like the one described here. These kinds of activities provide a common space where robotics and other kinds of technology, science content knowledge, imagination, and collaborative spirits can converge to work toward a common goal. After all, is this not what doing real science is all about? Why, then, should we not take steps to provide more opportunities for all students to engage in meaningful and collaborative learning?

NOTES

1. Generous grants from the National Science Foundation supported the Maxima and I2TechSciE Projects (Grants #9906339 and #0306156, respectively). The opinions and findings presented here, however, are those of the author only, and they do not represent the views of the funding agency.
2. More information on female astronauts can be found at NASA Quest (n.d.).

REFERENCES

American Educational Research Association. 2004. Closing the Gap: High Achievement for Students of Color. *Research Points* 2 (3): 1–4. www.aera.net.
Apple Computers. 2002. *The Impact of Technology on Student Achievement: A Summary of Research Findings on Technology's Impact in the Classroom.* www.apple.com/education.research.
CEO Forum on Education and Technology. 2001. *Year 4 Report: Key Building Blocks for Student Achievement in the 21st Century.* Washington, DC: Author.
Corbett, Dick, Bruce Wilson, and Belinda Williams. 2005. No Choice but Success. *Educational Leadership* 62 (6): 8–12.
Maxima. 2004. Welcome to the Maxima Project. http://edweb.sdsu.edu/maxima/.
NASA Quest. N.d. Welcome to the Women of NASA Content Area of NASA Quest. http://questdb.arc.nasa.gov/content_search_women.htm.
Newmann, Fred M., Anthony S. Bryk, and Jenny K. Nagaoka. 2001. *Authentic Intellectual Work and Standardized Tests: Conflict or Coexistence?* Chicago: Consortium on Chicago Research.
Pasco. 1996–2005. Innovative Solutions for Science Learning. www.pasco.com.
Robotics and Things. N.d. Robotics and Things: Robotic Creatures and Robotic Vehicles. www.roboticsandthings.com.
Rodriguez, Alberto J. 2004. *Turning Despondency into Hope: Charting New Paths to Improve Students' Achievement and Participation in Science Education.* Southeast Eisenhower Regional Consortium for Mathematics and Science Education. www.serve.org.
Rodriguez, Alberto J., and Cathy Zozakiewicz. 2005. "Using Sociotransformative Constructivism (sTc) to Unearth Gender Identity Discourses in Upper Elementary

Schools." *Penn GSE Perspectives on Urban Education* 3 (2). www.urbanedjournal
.org/index.html.

Rodriguez, Alberto J. (principal investigator), Cathy Zozakiewicz, and Randy Yerrick.
2005. Improving the Participation and Achievement of Students in Diverse
Schools by Enhancing Teacher Professional Development in Science and
Learning Technologies. http://edweb.sdsu.edu/i2techscie.

Vernier. N.d. Vernier Science & Technology. www.vernier.com.

Wenglinsky, Harold. 2000. *How Teaching Matters: Bringing the Classroom Back into
Discussions of Teacher Quality.* Princeton, NJ: Educational Testing Service.
www.ets.org/research/pic.

X10. 1997–2006. Main page. www.x10.com.

ADDITIONAL RESOURCES

Suggested Readings

Branwyn, G. 2004. *Absolute Beginner's Guide to Building Robots.* Indianapolis, IN: Que
Publishing.

Druin, A., and J. Hendler. 2000. *Robots for Kids: Exploring New Technologies for
Learning.* San Diego, CA: Academic Press.

Websites

Another great source of information on learning technologies: www.ceoforum.org
Great source of information on learning technologies: www.edutopia.org
I2TechSciE Project: http://edweb.sdsu.edu/i2techscie
Maxima project: http://edweb.sdsu.edu/maxima/
Women in NASA: http://questdb.arc.nasa.gov/content_search_women.htm

Sources for Equipment

Magnetometer and other computer probes: www.pasco.com
Robotix components: www.roboticsandthings.com
Temperature and other computer probes: www.vernier.com
Wireless camera and other equipment: www.x10.com

PART 5

NEW ROLES FOR TEACHERS AND STUDENTS

26

Teachers as Researchers

Sonya N. Martin

Teacher participation in educational research is not new. However, teachers' roles have been largely passive in that their cooperation with researchers, administrators, and policy makers has often relegated them to activities such as observing, testing, and implementing externally developed educational initiatives. Teachers have seldom been supported in investigating their own questions about their practices or their students' practices. Involving students as active members of classroom research is even rarer. At a time when educational research is being criticized for its distance from the classroom and its inability to effect change, providing teachers and students with the means to engage in research on their own classroom practices is relatively novel. This chapter provides insights into a process that brought research directly into my classroom, enabling me to make changes in my teaching while expanding learning opportunities for my students.

A DISCONNECT BETWEEN THEORY AND PRACTICE

Traditional classroom structures do not provide students and teachers with opportunities (or inclination) to discuss the roles and goals of students and teachers in the class—much less their interpersonal relationships. This is especially true in crowded urban schools, where teachers may encounter over 150 students per day. The sheer number of students alone provides an extra hurdle that would make these types of conversations improbable. This means that teachers are generally uninformed about a wide range of topics that can affect student participation and achievement in their classroom. As an urban educator, I faced many constraints like these. Stepping into the classroom as a new teacher in a new school, I had an endless number of responsibilities to learn to manage on top of figuring out how to effectively "teach" the content for each assigned class. As a result, I often felt disconnected from my students, making it difficult to anticipate their needs as learners and impossible to adapt my teaching practices to better support them. Thus, even while I realized that I needed to make changes to improve learning in my classroom, I had no means to support such an endeavor.

In my teacher education courses, I was introduced to terms such as *teacher-researcher, action research,* and *reflective practitioner.* I was urged to reflect on my teaching practices, to *journal* about my interactions with students, and to pose questions and seek answers concerning instruction, curriculum choices, and the learning environment. In theory I was prepared to do all of these things, but in practice it seemed an impossible task. I could find no way to bridge the expanse between what I was introduced to in educational theory and what was realistically possible—until I was introduced to a model of teacher research that provided me with the tools I needed to co-construct that bridge with my students. The collaborative ethnographic research I describe below provided my students and me with structures to support new ways of interacting with one another, expand our roles, and position us as agents of change in *our* classroom.

RESEARCHERS—NOT RESEARCHED

Traditional ethnography relies heavily on thick description of culture and social interactions in an attempt to construct understanding, but such accounts rarely include the "subjects" of study in the research process itself. Angela Calabrese Barton (2001, 912) refers to this research methodology as *participatory pedagogy,* stating that "research must be a fully reflexive process" in that the research process becomes "research *with* rather than research on or even *for.*" When invited to become a member of a study funded by the National Science Foundation to study science teaching and learning in urban classrooms, I was wary about inviting outsiders into my classroom because my previous experiences with educational research involved teachers as subjects rather than participants. However, I was optimistic that this research would be different from more traditional research arrangements because the focus was on improving *my* practices in *my* classroom. Collaboration was not limited to teachers and university researchers, but included students and their perspectives about what it was like to be a student in *my* classroom. The research methodology was specifically designed with these issues in mind, leaving space for teacher researchers, university researchers, and student researchers to develop their own interests and address their own emerging research questions. As a result, my students and I formed our own research questions, collected data to answer them, and analyzed our data using the methodological and theoretical frameworks introduced to us by our university researcher. Our involvement promised immediate, tangible benefits for us as we assumed the roles of researchers—rather than the researched.

Before my participation in this research, I had given very little thought as to how I interacted with my students or even why I enacted some teaching practices and not others. I relied on a trial-and-error method of teaching, rarely reflecting on why certain practices were more effective than others, and only recognizing that "some things worked and others didn't." This is the limit of *reflection* as provided by many action research models, and very few of these models include students in the process. My involvement with students as researchers provided me with a window into their lives, encouraging them to share with me their personal experiences, opening a door that was closed to me as a "traditional teacher." Our coparticipation in this process provided students with unique opportunities as well—encouraging them to reflect on their roles in the classroom and on the teachers' roles. The inclusion of multiple perspectives in the design of this research allowed us to go beyond "what does and does not work" in an effort to understand "how and why *our* classroom works the way it does." This understanding was significant in that it enabled us to begin to identify beneficial practices to be supported and strengthened as well as problems to be resolved. No longer would I be the only one responsible for "troubleshooting" problems in the classroom or for applying self-developed "remedies for change." By sharing this responsibility with my students, I embarked on what seemed a more manageable course of classroom research—with others.

TOOLS FOR CHANGE

Several aspects of this research model were influential in providing my students and me with opportunities for sharing our individual perspectives about our experiences *in* class and *with* science. The two most salient features included viewing video recordings of our classroom and participating in cogenerative dialogues. Often, the two were combined to permit our collective analysis of video clips to identify topics for discussion in cogenerative dialogues (LaVan 2004), which are discussions between stakeholders that examine shared events and experiences. Described by Wolff-Michael Roth and Kenneth Tobin (2002, 252), cogenerative dialogue enables participants to "use current understandings to describe what has happened, identify problems, articulate problems in terms of contradictions, and frame options that provide us with new and increased choices for enacting teaching and learning." The notion of shared responsibility is central to these discussions, as participants reflect on shared experiences, power relationships, and the differing roles and perspectives of all those involved. Shared perspectives are used to inform the emerging understandings of classroom interactions, the quality of these interactions, participant practices, and how these patterns of interactions contribute to the collective activity of teaching and learning science.

Videotape provided an effective means for us to reflect on our classroom practices, and cogenerative dialogue provided social spaces for us to consider our practices with others. Video and audio analysis added a layer of complexity to the process of reflection that is often lacking in action research as commonly practiced by teacher-researchers. By capturing classroom interactions on video or audio, teachers and students have access to a record of the classroom that is not subject to the reconstruction of a moment based on individual or collective "memories of events." Video analyses of classroom interactions inform teachers and students of what happens in their classroom and help them to recognize the ways in which their unconscious practices shape the learning environment. Cogenerative dialogues complement this process by enabling us to talk about issues in the classroom in ways that are both constructive and nonconfrontational, identifying and discussing issues of concern, resolving contradictions, and creating options for action to produce desirable changes. In this way, cogenerative dialogues created social spaces for me to interact with students, creating social interaction patterns that differed from those associated with traditional teacher and student roles. Conversations regarding video of our classroom interactions changed as our social bonds strengthened. This process took time because we had to develop mutual trust, get to know one another, learn to become critical of our actions and interactions (using video and audio analysis, interviews, journals, and collaborative research meetings), and then bring all of this back to the classroom to inform our changing practices.

BUILDING BRIDGES IN THE CLASSROOM

As my students and I had more opportunities to interact in cogenerative dialogues with one another, we developed a better sense of how to "talk to one another with respect." Interacting with adults on a regular basis in this way allowed my students to develop new ways of speaking to adults and being heard by them. Similarly, I was learning how better to listen to my students and make myself heard. We explicitly discussed the ways in which we communicated, examining the language we used and the manner in which we approached one another. These were important skills for us to develop because it provided me with different tools for interacting with students and it helped students to improve their interactions with teachers and other adults. This increased dialogue provided opportunities for me to communicate more fully my expectations for their learning while simultaneously positioning them as more equal stakeholders in their own learning.

Some examples of issues we addressed included factors that prevented students from coming prepared to class, including inability to juggle numerous outside commitments (e.g., sports, after-school jobs, and other coursework) and the nature of assignments (perceived relevance or importance of assignment and/or access to necessary resources to complete assignments). By exposing these issues, we began to address the structural changes necessary to increase student participation. As students voiced their beliefs about class and about what best supported their learning, they found they were no longer isolated in their struggle to learn chemistry. They began to identify strategies they felt supported their learning and shared them with their peers and me.

As I learned more about my students' lives through our research, I began to understand the ways in which seemingly inconsequential actions on my part could have substantial effects on their lives. Altering due dates for projects or postponing exams, I learned, could seriously jeopardize a student's ability to perform well in my course if that student's living arrangements (living in a shelter, traveling between two parents, or even observance of religious holidays) necessitated a schedule that remained fixed and consistent. Inconsistency on my part sometimes placed students at a disadvantage in that they were unable to complete assignments that required resources they could not access. This example makes salient the fact that a teacher's classroom practices need not be exceptionally cruel, overtly racist, or inherently uncaring to have a negative impact on student participation, achievement, or self-esteem. However without video analysis of classroom interactions and co-generative dialogues, most of these realities would have remained hidden.

THE ROLE OF COLLABORATIVE REFLECTION
IN CLASSROOM RESEARCH

The collaborative reflective process was greatly enhanced by the addition of video and cogenerative dialogues with my students. Both provided us with a more complex picture of what was (and was not) occurring in the classroom. This triangulation of classroom perspectives not only enhanced my understanding of the classroom, but also benefited students who developed a greater appreciation of the roles played by both the teacher and the student in the teaching and learning equation. Most students stated that other than recognizing that they seemed to perform better in some classes and with some teachers as opposed to others, they had not given much thought as to why this was so. However, this research provided students with the structure they needed to examine critically their own practices and relationships with others. An example of this comes from a transcript of a conversation with a student researcher as she reflected on the need for more student/teacher research alliances.

> Before [this experience], I really only thought of teachers as being good or bad at teaching or me being good or bad in a subject. Now I think more about what makes the teacher good or bad or makes me perform good or bad in class. I now know there is more to it—including how teachers and students feel about one another, about the subject, and how and what they do in a class—like their actions. Thinking and talking about this stuff makes it clear that teachers and students are not just good or bad at a subject—there are things we could look at,

maybe do a different way, so we could improve it. (Danni electronic correspondence, March 2003)

Participating in this collaborative reflection enabled participants to become more aware of their roles in the classroom. Recognition of these roles allows teachers and students to begin to consider their individual responsibilities for contributing positively to their learning and to the learning of others. Our experiences in cogenerative dialogues, over time, provided us with new ways of talking to and listening to one another. This form of collaborative, reflective research provides teachers and students a means to communicate with one another that helps bridge the gaps caused by differences in age, race, gender, and class. There is a saying that "once you look at things in a different way, you start to see things differently." The process of reflection and research that my students participated in enabled us to "start to see things differently" because it provided a structure from which to "look at things in a different way."

FROM RESEARCHED TO RESEARCHER

At a time when teachers are being held "accountable" for their students' failures, it is more difficult than ever for a teacher (especially urban teachers) to engage in an open dialogue with others about his or her weaknesses—effectively preventing any possibility for dialogues about how to make change. Engaging in this research was a risky decision for me to make, but it was empowering because it meant that as a teacher I could begin to identify (and eventually own) my mistakes. I started developing the courage to admit my shortcomings because I felt empowered by the belief that, with the help of others, I could implement real change. Just as I expected my students to learn from their mistakes, I started to change the expectations I held for myself as a teacher. No longer engaging in "teacher talk," the conversations I now had with other teachers (and students) about my classroom seemed more legitimate. Reflecting upon my practices with others, I started to feel empowered by my role as a teacher-researcher. No longer afraid that someone would find out I was not perfect, I began to share with parents, students, and other teachers by saying things like "the research we've been doing in my classroom shows..." or "in researching my teaching practices, I've noticed..." because I now felt a sense of responsibility to acknowledge my faults knowing that I could work collectively with others to resolve contradictions.

Making the transition from what Schön (1983) refers to as the "defensive expert" to that of a "reflective practitioner" requires a process of self-examination that is difficult and sometimes painful. Describing her experiences as a teacher-researcher of her own classroom teaching practices, Wendy Schoener (Ulichny and Schoener 1996, 523) states that she learned that she cannot "divorce who I am from how I teach" and that while she will always face difficulties in her teaching, she now recognizes that the "greatest possibilities lie in paying attention to them." These words resonate deeply with my own experiences as a teacher-researcher. In fact, they form the kernel of my concluding thoughts for why this collaborative research is so transformative for teachers and their students. The potential to transform the classroom lies in knowing oneself as a teacher and/or learner, and only by collectively seeking to expose and examine the issues associated with the process of teaching and learning can contradictions be resolved to afford greater possibilities for all classroom participants.

REFERENCES

Barton, Angela C. 2001. Science Education in Urban Settings: Seeking New Ways of Praxis through Critical Ethnography. *Journal of Research in Science Teaching* 38:899–917.

LaVan, Sarah-Kate. 2004. Cogenerating Fluency in Urban Science Classrooms. Ph.D. diss., University of Pennsylvania.

Roth, Wolff-Michael, and Kenneth Tobin. 2002. *At the Elbows of Another: Learning to Teach through Coteaching*. New York: Peter Lang.

Schön, Donald. 1983. *The Reflective Practitioner*. New York: Basic Books.

Ulichny, Polly, and Wendy Schoener. 1996. Teacher-Researcher Collaboration from Two Perspectives. *Harvard Educational Review* 66:496–528.

ADDITIONAL RESOURCES

Elmesky, Rowhea, and Kenneth Tobin. 2005. Expanding Our Understandings of Urban Science Education by Expanding the Roles of Students as Researchers. *Journal of Research in Science Teaching* 42:807–28.

LaVan, Sarah-Kate, and Jennifer Beers. 2005. The Role of Cogenerative Dialogue in Learning to Teach and Transforming Learning Environments. In *Improving Urban Science Education: New Roles for Teachers, Students, and Researchers*, new ed. by Kenneth Tobin, Rowhea Elmesky, and Gale Seiler, 167–85. New York: Rowman & Littlefield.

Martin, Sonya N. 2005. The Social and Cultural Dimensions of Successful Teaching and Learning of Science in an Urban High School. Ph.D. diss., Curtin University, Perth, Australia.

Wassell, Beth. 2004. On Becoming an Urban Teacher: Exploring Agency through the Journey from Student to First Year Practitioner. Ph.D. diss., University of Pennsylvania, Philadelphia.

27

Choreographing Teaching: Coteaching with Special Education/ Inclusion Teachers in Science Classrooms

Susan Gleason, Melissa Fennemore, and Kathryn Scantlebury

Sue, an experienced high school chemistry teacher, is in the front of the classroom teaching chemical bonding to college-prep/inclusion sophomore students, and it is not going well. The topic is difficult, and the students are easily distracted. Melissa, a second-year inclusion science teacher with a chemistry degree, while helping two students, begins to understand the students' confusion. Melissa moves to the front of the room, interjects a clarifying analogy, and redirects the lesson. Seamlessly, Sue takes over Melissa's role of working with individual students. They are coteaching.

On another day, the lesson plans call for Sue to finish teaching atomic structure. Two students who have been absent for several days do not understand the material. Another student already has her hand in the air. As Sue begins, Melissa takes the two returning students to the back of the classroom to review the material. Sue and Melissa are coteaching.

The two vignettes describe how experienced teachers utilize a model of coteaching to improve students' science learning in inclusion classes. The student and his or her learning comprise the focal point in coteaching. Wolff-Michael Roth and Kenneth Tobin (2002) explained how coteaching provides teachers with shared teaching experiences that they can then reflect upon and analyze to improve their teaching and students' learning. When coteaching is used as a form of professional development, teachers are active participants. Often, in-school/on-site professional development programs are passive experiences for teachers, who may have difficulty translating the ideas to improve their classroom practice. Frequently, there is a disconnection between ideas presented in professional development seminars and the teachers' classroom situations and experiences. Coteaching provides a structure for teachers to reflect upon their teaching, coplan lessons, and explore strategies and new ideas that will enhance students' learning.

Successful coteaching exists in a dialectical relationship with co-respect, co-responsibility, coplanning, and cogenerative dialogues. The foundational "co" of these is *co-respect*. The teachers must view each other as peers, each person providing insight and knowledge that will improve his or her teaching and enhance students' learning. That knowledge and insight vary with each teacher. In the examples above, Sue has eighteen years' experience of science teaching. Melissa, the inclusion coteacher, brought knowledge of inclusion students and strategies to Sue's classes. Sue and Melissa often coteach using a conversational style by including the students in the discussion. They also use each other as resources and bring in other resources such as students' science knowledge. During the classroom discussions, they ask questions of each other and show the students that they are willing to learn from each other.

When coteaching, Sue and Melissa are equally responsible for all students' learning, and successful coteaching is dependent upon teachers assuming co-responsibility. Thus, when Melissa steps in to clarify a concept or monitor the students' interactions with Sue, she is actively sharing the teaching space. In coteaching, while one teacher is providing direct instruction, the other can scan students' faces and body language to ascertain if they understand the explanation. The teachers are co-responsible for the students' learning.

Co-responsibility occurs when each coteacher assumes responsibility for all aspects of the classroom: the instruction, the students, and the teaching and learning outcomes. When Sue and Melissa coteach science, they have to remain focused on the unfolding lesson and classroom events. At any given time, the teaching activity could vary. Sometimes the two teachers are "tag-teaming" the lesson with one supporting the other with additional instruction or writing on the blackboard, or working separately with a student or a group of students. Coteaching is not haphazard or spontaneous, but is the result of coplanning.

Coplanning requires the participation and involvement of all coteachers. The act is labor-intensive because each teacher reflects upon prior lessons, program objectives, and goals. They also must relate the learning to standards and provide input on artifacts such as worksheets, quizzes, or laboratories. During the planning, coteachers begin choreographing lessons and decide which teacher will lead and who will provide individual instruction or manage the administrative tasks such as taking attendance, entering grades, or grading papers. Coplanning is an excellent professional development activity because during this time, teachers share ideas, use past experiences, discover new activities, and develop an understanding of student needs. The follow-up, another professional development experience for teachers, to a coplanned and cotaught lesson is a cogenerative dialogue.

Cogenerative dialogues are open discussions where all participants' opinions have equal value. Cogenerative dialogues can take many forms, and one is a debriefing session focused on the implementation of an activity, a lesson, or an assessment. Cogenerative dialogues provide opportunities to reflect on teaching praxis. How did the coteachers move from a coplanned lesson to its enactment? What strategies were successful? Which students need further assistance? How will the curriculum unfold in future lessons? Cogenerative dialogues are not forums for assigning blame for an unsuccessful lesson but an opportunity to critically review teaching practice and strategize ways in which all coteachers may improve. It is a support group of peers. Melissa reflects, "We get to discuss things and get to say what was good enough about that. And we get to go over things, and say, 'Well, that really stunk. What would you do to make it better?' And go over that." Unlike Lesson Study, where teachers observe a colleague and offer critique, coteaching provides teachers an insider's perspective to the teaching and learning (Fernandez, Cannon, and Chokshi 2003). By participating in the lesson, teachers develop an understanding and appreciation of the nuances in teaching a particular topic, such as atomic structure. Throughout planning, instruction, and the cogenerative dialogue, coteachers are collectively responsible for the lesson with their peers.

Even in the cotaught, coplanned class, lessons are not always successful. Coteachers can become supportive resources, offer alternative explanations, clarify activity instructions, or provide individual instruction to demanding student(s). We have found that coteaching resources are akin to having at your side a private rescue squad. With your pooled knowledge, pooled experience, and extra hands, lessons are saved, quickly restructured to meet students' needs or modified for diverse learning situations.

When we began using coteaching in inclusions classes, Sue noted,

> I am particularly excited about [coteaching's] application with inclusion classes. I learned from my special education inclusion partner Melissa much about teaching special education students, something I had never done before and was apprehensive about doing. With my teaching experience I was able to show her my types of activities she was not familiar with. Together we adopted a common approach to situations, both with the class as a whole and with individual students. We used our individual strengths and our combined strength to better address student needs. We learned from each other and the students.

Melissa also worked with three other science teachers, and as a beginning teacher, her professional development experiences come from working with her colleagues. Her coteaching experiences provide an opportunity for her to develop her teaching craft at the elbow of others. And they in turn have a colleague with whom they can collaborate who understands the students and the teaching environment.

The *National Science Education Standards* (National Research Council 1996) encourages teachers to be collaborative and reflective. Coteaching is collaborative: two or more teachers work together to educate their students. To be successful, the teachers must coplan and be co-responsible. Coteaching sets up a structure for teachers to be more reflective about their practice, to participate in cogenerative dialogues, and to have a peer involved with their teaching praxis. When two or more people discuss a lesson, each teacher needs to know how the lesson relates to standards, and what assessments and evaluations will be used, and the cooperation and collaboration build co-respect between peers. Coteaching provides teachers the space to articulate and share their teaching knowledge, and, when linking theory to practice, colleagues with whom to share that experience. Overall, through coteaching, coplanning, accepting co-responsibility, and showing co-respect, the teachers improved their teaching through reflective praxis.

REFERENCES

Fernandez, Clea, Joanna Cannon, and Sonal Chokshi. 2003. A U.S.-Japan Lesson Study Collaboration Reveals Critical Lenses for Examining Practice. *Teaching and Teacher Education* 19:171–85.

National Research Council. 1996. *National Science Education Standards*. Washington, DC: National Academy Press.

Roth, Wolff-Michael, and Kenneth Tobin. 2002. *At the Elbow of Another: Learning to Teach by Coteaching*. New York: Peter Lang.

28

Theater of the Oppressed Pedagogy: Fostering Critically Reflective Science Teachers

Deborah J. Tippins and Foram Bhukhanwala

Today there is widespread agreement that an important goal of science teacher education is to prepare teachers as critically reflective practitioners. For many years, traditional conceptions of learning to teach were framed as a matter of translating learning theories to practice. This technical/rational approach to teacher education, what Jean Clandinin (1995) calls "the sacred theory-practice story," has prevailed for decades, characterized by a collection of discrete and separate courses where theory has very little real connection to practice. In recent years, with the emergence of professional development schools and other more inquiry-oriented models of teacher preparation, reflection has become central to integrating theory with practice.

The origin of reflection in teacher education can be traced back to John Dewey (1933) and his emphasis on conscious, systematic, and "practical deliberation" as a way of enabling teachers to analyze, discuss, evaluate, and change problems in their own teaching practice. Donald Schön (1983), building on the earlier work of Dewey, emphasized the importance of reflection-in-action as an interactive way of framing a dilemma in the context of practice and reflection-on-action as a type of deliberative inquiry undertaken to change future actions. More recently, the term *critical reflection* has been used to describe inquiry, which challenges traditions and conventions that have been dominant in science teacher education in ways that reveal ethical, moral, and ideological aspects of teacher thinking and action. Since the early 1980s, with the recognition that reflection is integral to viewing learning to teach as a "problematic" endeavor, research into strategies for promoting reflection in prospective and practicing science teachers has become prevalent. Metaphors, cases, journals, learning maps, photo essays, portfolios, and proverbs are just some of the strategies that have been used to engage teachers in a cycle of systematic reflection central to their practice. Schön emphasized that different strategies and conceptions of reflection should lead to a restructuring or "reframing" of a teaching experience and ultimately contribute to new insights. Pedagogy and Theater of the Oppressed is a reflection strategy that uses creative expression to expand the boundaries of how teachers come to view the teaching of science as a practice that cannot be left unchallenged.

PEDAGOGY AND THEATER OF THE OPPRESSED

This form of theater is a powerful visual and mental tool, which was developed in Brazil by Augusto Boal—an actor turned social activist. It has been internationally practiced and adapted to working with different groups like youth, women, people in prison, refugees in camps, teachers, and students. It is intended to help people, including teachers and students, recognize oppressions in their lives and then find ways to overcome these oppressions (Boal 2003). In a safe environment, by participating in various theater games, teachers can develop consciousness about themselves, their actions, and the effects their actions can have on others. In brief, this form of theater can be used with a group of teachers or students who feel oppressed and powerless to change the already existing institutionalized way of doing things.

Boal's work was influenced by Paulo Freire and the liberation pedagogy he developed. Freire viewed education as "problem posing" and believed the aim of education was to develop critical consciousness. Critical consciousness is an individual's ability to "perceive social, political and economic contradictions, take action against the oppressive elements of reality" (Freire 2001, 35), and recognize and articulate personally challenging experiences. Freire believed that by observing what we see, listening to what we hear, and feeling what we touch, we come to know the world in which we live. The development of a critical consciousness is likely to prompt teachers to participate in situations where they can promote equality and social justice, as they come to realize that science learning does not take place in an economic and social vacuum. A critically reflective teacher carries on a long-running dialogue with self, questioning the forces that have shaped schools and society, and considering the circumstances that give rise to oppression rather than individual stories of suffering.

Image Theater is a "tool" for helping teachers make sense of experience and, guided by inner conversations, reframe personal constructions of teaching. It consists of a series of games and exercises where teacher or student "actors" create images by molding their bodies as if they were clay in order to express an image in a realistic or symbolic form. Language consists of images that are not usually revealed in speech; they are a language in themselves and cannot be translated into words. To read an image, one needs to "feel those images, to let our memories and imaginations wander: *the meaning of an image is the image itself*" (Boal 2003, 175 italics in original). During a game or exercise, other teacher observers orally express the feelings that are evoked or the thoughts that are created through their imagination and participation in Image Theater. These multiple reflections bring alive the hidden aspects of the image to the creator, who then has the freedom to choose the meaning he or she wants to make through the process. In this sense, Image Theater can be used as a liberatory pedagogy. Oppressions are often "kinesthetically inscribed"; thus, for an individual to feel liberated, the response, too, must be kinesthetic. Therefore, an individual is likely to feel liberated when he or she modifies an image of powerlessness to create a more liberated posture. Through creating and reflecting on these images, teachers begin to consider how their beliefs and actions are influenced by social, political, and economic forces. By engaging in self-reflection and dialogue, they recognize ways in which these forces may limit their imagination, and begin to generate alternative visions of what science teaching and learning might be like. We constructed the following vignettes to illustrate how prospective teachers might make sense of their world and a view of science through Image Theater.

VIGNETTE #1: A VIEW OF OUR SCIENCE CLASSROOMS

The prospective teachers had returned to their science education course after spending a month "in the field" experiencing the lifeworld of middle school science classrooms. The class was filled with chatter, laughter, and emotions as they talked about their experiences in the field. I divided the prospective teachers into four small groups and asked them to share their experiences with peers. Quickly, they were engaged in animated discussion as they talked about their lived experiences with science teaching and learning. I was struck by the depth of their conversation and the level of engagement. I was pleased to see the excitement building in my class, and the enthusiasm with which students talked and listened to one another. It affirmed my belief that students feel engaged and motivated when they have the opportunity to share and reflect on lived experiences.

I asked the group to come to the center of the room and form a circle. Once they had formed a circle, I asked each to think of a particular science class and form an image to express his or her thoughts and feeling. Some of the images created were of students sleeping, looking bored, looking away, writing, and reading. Other images were created where the teacher questioned, was nurturing, or pointed to a distant object. One prospective teacher shaped her body in the form of a question mark and had a quizzical expression on her face. Once the group had formed a collection of images, individuals stepped back, one at a time, to view the images that were created by their classmates.

VIGNETTE #2: IMAGINING OUR FUTURE
SCIENCE CLASSROOMS

We sat in a circle and talked about the images they had created and what these images could mean for science teaching and learning. The prospective teachers immediately started talking, but the tone of their conversation had changed. They started sharing about experiences that were disturbing, raising questions such as "What do you mean when mentor teachers will not allow you to try out a pedagogy that may seem different than theirs?" "How should you respond when other teachers say that instructional time must only be used to prepare students for the standardized test?" "What counts as knowledge in science classrooms, and who decides?" and "Why are children not interested in learning science?" The questions were many, each reflecting some form of oppression; and as each question emerged, I heard gasps, whispers, and nervous laughter. Finally, one prospective teacher asked aloud, "What can be done about our questions?" There was a long pause. . . . The group of prospective teachers had begun to think, wonder, and look for possibilities. Viewing this as a teachable moment, I invited them back to form a new image to consider what they would like their science classroom to be like.

Time ran out, and the class period was over. But that day, prospective teachers went home with a question, a doubt. They wondered about teaching, about their relations with their mentor teacher, and about ways in which their voices, and those of their students, could be heard. Image Theater helped to "problematize" teaching by enabling a group of prospective science teachers to develop insights that could help them recognize and challenge oppression as they sought ways to create their own liberation and actively claim their education.

Image Theater can be used throughout teaching as a vehicle for teachers to give meaning to their experiences through reflection. Moral dilemmas are likely to surface as teachers participate in Boalian Theater. By creating images, the unspoken becomes enacted, thereby creating a platform for furthering actions. Specifically, by considering multiple interpretations of actions and events, teachers can expand their understandings of what it means to teach and learn in a pluralistic society. In the process, they may question and move beyond conventional interpretations of practice to consider alternative ways of building socially responsive science classrooms.

REFERENCES

Boal, Augusto. 2003. *Games for Actors and Non-Actors.* New York: Routledge.
Clandinin, Jean D. 1995. Still Learning to Teach. In *Teachers Who Teach Teachers,* edited by Tom Russell and Fred Korthagen, 25–31. London: Falmer.
Dewey, John. 1933. *How We Think.* New York: Heath.
Freire, Paulo. 2001. *The Pedagogy of the Oppressed.* New York: Continuum.

Schön, Donald. 1983. *The Reflective Practitioner: How Professionals Think in Action*. New York: Basic Books.

ADDITIONAL RESOURCES

Bullough, Robert V., and Andrew Gitlin. 1981. Educative Communities and the Development of the Reflective Practitioner. In *Issues and Practices in Inquiry-oriented Teacher Education*, edited by Robert Tabachnich and Kenneth Zeichner, 33–55. Bristol, PA: Falmer.

Creel, Gill, Michael Kunhe, and Maddy Riggle. 2000. See the Boal, Be the Boal: Theatre of the Oppressed and Composition Courses. *Teaching English in the Two Year College* 28:141–56.

Day, Laura. 2002. Putting Yourself in Other People's Shoes: The Use of Forum Theatre to Explore Refugee and Homeless Issues in Schools. *Journal of Moral Education* 31 (11): 21–34.

Nichols, Sharon E., and Deborah J. Tippins. 1997. A Toolkit for Developing Critically Reflective Science Teachers. *Journal of Science Teacher Education* 8:77–106.

Zeichner, Kenneth, and Jennifer M. Gore. 1990. Teacher Socialization. In *Handbook of Research on Teacher Education*, edited by W. Robert Houston, 329–48. New York: Macmillan.

29

Constructing Knowledge about Electricity, Motors, and Magnets outside of the Box

Judith A. McGonigal

Science in the elementary classroom is presently defined by the culture of standards. The scope and sequence of science lessons are not driven by students' questions or interests, or by a teacher's enthusiasm or expertise. Instead, every decision focuses on aligning a classroom's elementary science curriculum with national and state standards.

One consequence of this focus is the prepackaged science lesson. A teacher is provided a manual of sixteen lessons, along with a kit of materials to support each lesson. These lessons-in-a-box are designed to help a school district meet the national and state standards. However, they do not always support—and, in some cases, can actually undermine—scientific learning. This personal reflection about teaching elementary science identifies the potential pitfalls of using these prepackaged lessons and describes how teachers can "think outside the box" to maximize students' learning.

FOLLOWING THE FORMULA

Classroom teachers often lack sufficient understanding of the science behind each prepackaged lesson. The teacher follows the directions in the manual like a cookbook, hoping that the standards-based teaching recipes will result in learning. However, it is not uncommon for students to follow the prescribed investigations and fail to obtain the results described in the manual. The elementary classroom teacher often lacks the confidence, and the scientific knowledge, to explain or investigate the discrepancies. In the next round of teaching the kit, the teacher simply skips those lessons that seemed to "not work." Eventually, the science kit becomes a box of watered-down activities, and the lessons lack the rigor envisioned by the developers of the standards and the authors of the science kit.

WHAT IS AN ALTERNATIVE RESPONSE?

Elementary grade teachers can become co-learners with their students, engaging in a shared inquiry about the content focus of the science kit. The kit becomes a starting point for—but not a limit to—classroom scientific investigation.

I took this approach when my school district curriculum required that I use a science kit and manual, *Magnets and Motors* (National Science Resource Center 1991), in my fifth-grade classroom. This kit was selected to fulfill the National Science Education Content Standard B, "All students should develop an understanding of electricity, and magnetism" (National Research Council 1996, 123), and the New Jersey Science Content Standard 5.7 (Physics), "By the end of Grade 4, students will recognize that some forces can act at a distance (magnetism and electricity)," and "By the end of Grade 6, students will design an electric circuit to investigate the behavior of a system" (New Jersey Department of Education 2004).

Because I had little knowledge of magnets and motors, the teacher's manual became my road map. However, when I reached "Lesson 7: Creating Magnetism through Electricity," I lost my way. I followed the lesson, but I did not understand how to direct my students' learning. The lesson required students to use "a magnetic compass to detect whether current was flowing through the wires of a circuit" (National Science Resource Center 1991, 45). I was amazed and confused by what the students observed: the compass needle was affected by the electric current in a closed circuit. Because I had no concept of the relationship between magnetism and electricity, I could not effectively teach this lesson.

What was I to do? I was an elementary grade teacher in a self-contained classroom. My district had no science curriculum supervisor from whom I could seek help. For two months, I abandoned the kit and stopped teaching science to my class. However, I was committed to finding a way to teach this science content effectively.

SEARCHING TO LEARN THE CONTENT

To increase my knowledge of electricity and magnets, I searched the Internet for information. I stumbled upon a website inviting teachers of grades 4–8 to participate in a Hands-on/Virtual Inquiry Physics course about electricity and magnetism. I enrolled. When the white box of materials arrived at my home, with another manual (Straley and Shafer 2003), I decided to ask my fifth-grade students to help me take this course. Six students offered to join a science club that met after school for six weeks.

We explored the materials in the new box, following the format of some of the lessons in the manual, and designing our own investigations about electricity and magnets. For example, we worked with siphons to study the flow of fluids; explored electrical devices, including a switch and a light-emitting diode; and "dissected" a light bulb and a motor. We co-constructed an elementary understanding of electricity, magnetism, and motors though this combination of manual-directed and self-directed inquiry.

RETURNING TO THE KIT WITH NEW UNDERSTANDING

With this new understanding, I was able to return to the science kit and continue the prescribed lessons with confidence. When my fifth-grade class reached Lesson 9 in *Magnets and Motors*, the students were invited to design and carry out their own controlled experiment to determine the effects of different variables on the strength of an electromagnet.

Students chose their own variables to manipulate. Some teams decided to vary the number of turns of wire around a metal core. Other teams, wanting to see what would happen if they manipulated the size of the core, sent me to Home Depot to purchase bolts of varying diameters and lengths. One team decided to investigate whether changing the location of an electromagnet in the classroom would affect its strength. This team observed that room location did seem to influence the magnet's strength. Although I thought that the students' conclusion was wrong, I allowed them to report their evidence to our community of young scientists.

CONTINUING TO LEARN OUTSIDE OF THE BOX

Four months later, I was asked to speak about science inquiry to a group of educators in New York City. In my presentation, I described how students can design and carry out their own science investigations. I used Lesson 9 of the *Magnets and Motors* kit—designing a controlled experiment—as an example.

When I invited questions, I was informed that one of the key authors of *Magnets and Motors* was in the audience. He explained that when the developers of the manual conducted field tests of the lessons, they had found that steel cabinets and desks in science labs could significantly affect a compass. He thought that my students' surprising finding warranted further investigation. He applauded the way students had been encouraged to design investigations based on their own questions about electromagnets.

Our curiosity to learn about magnets and motors was stimulated by the manual's prescribed lessons. However, our path to deeper understanding of the content was not contained in a box. By handling the materials and sharing in science discourse, we uncovered new questions to explore. When the science curriculum is defined by a lock-step series of standards-based lessons, teachers must be encouraged to provide room for students to identify their own questions and design additional investigations.

EXPANDING THE USE OF THE MATERIALS IN THE BOX

When I revisited this kit the next year, I was able to support my students' learning even when our investigations diverged from the expected results. When students failed to make a compass rotate using one electromagnet, as depicted in the manual, we problem solved as colearners. I remembered that our small science club, the prior year, had disassembled a toy motor from Radio Shack and had uncovered three electromagnets around the motor shaft. Based on this experience, I redefined the task of the lesson, and invited students to develop a way to rotate the compass using a system of electromagnets. The fifth graders worked in teams, and they successfully identified how to switch the power of each electromagnet on and off to make the compass needle spin like a motor.

The following week, I invited my students to use the materials of the kit to again go beyond the manual's content. I asked students to apply what they had learned to design a toy, using the magnets and motors provided in the kit. The students invented twenty-one different toys, including magnetic board games, motorized cars, and a battery-operated paddleboat that came with its own videotaped manual of operation.

Learning about electricity, magnets, and motors was now an exciting learning journey for my students and me. I learned to expand prescribed lessons and work outside the box of a standards-based manual by doing the following:

1. Finding additional investigations that honored the questions and confusions that emerged as cookbook experiments were carried out

2. Encouraging students to redefine the parameters of prescribed investigations

3. Problem solving with students, as a colearner, to design ways to accomplish tasks that appear not to work

4. Designing additional tasks that require the application of knowledge constructed in the kit lessons

It was important for me to understand that as a teacher, or a student, conducting science is an ever-expanding journey that cannot be contained in a box.

REFERENCES

National Research Council. 1996. *National Science Education Standards.* Washington, DC: National Academy Press.

National Science Resource Center. 1991. *Magnets and Motors,* teacher's guide. Burlington, NC: Carolina Biological Supply Company.

New Jersey Department of Education. 2004. New Jersey Core Curriculum Content Standards for Science. www.state.nj.us/njded/cccs/s5_science.htm#57/.

Straley, Joseph, and Sally A. Shafer. 2003. *Electricity and Magnetism 2003 Hands On/ Virtual Inquiry Physics for Teachers of Grades 4 through 8.* Lexington, KY: Shafer & Straley.

ADDITIONAL RESOURCES

Handwerker, Mark J. 2005. *Science Essentials: Lessons and Activities for Test Preparation,* elementary ed. San Francisco: Jossey-Bass.

Harlen, Wynne. 2001. *Primary Science: Taking the Plunge.* Portsmouth, NH: Heinemann.

Straley, Joseph. 2004. Online Courses and Professional Development in Physical Science. www.pa.uky.edu/sciworks/kntro.htm/.

30

Students as Researchers: Creating New Identities

Rowhea Elmesky

Schools have become places where students who exhibit practices that seem outside of the norms of the Eurocentric dominant culture are viewed as "other." Nowhere is this truer than in the neighborhood schools, with majority student populations of color, located in the largest urban centers. With teachers who are predominantly white and middle class, urban science classrooms are often sites of cultural conflict. Student practices that look, sound, or feel differently from those established by school norms become a source of frustration for teachers who try to control the behaviors. Teaching approaches then become about restraint rather than about learning; in science classrooms, that may mean shutting down the very practices that allow the students to think critically. When classrooms are conflicted, the emotional state is negative, activities of learning feel disjointed, and gaps of insurmountable distance seem to form not only between teachers and students who differ from one another but also between students and the subject matter of science. Consequently, marginalized students may not be able to develop strong identifications with school or with scientific practices.

In this chapter, I suggest that improved science teaching and learning are connected to the reduction of classroom cultural conflict. Specifically, gaps can decrease as we embrace methods for developing authentic understandings of the students we teach so as to positively inform our teaching practices. What teachers know about their students is very often determined by what others have said, the reputations that they have attained, and the larger images that are presented in society. When the students are of color and/or poor, these perceptions are commonly deficit in nature. Because racially margin-alized and economically disadvantaged children represent the most silenced groups in schools (and in society), it is essential that we consider the ways in which their disempowerment in academic settings can be transformed. In this chapter, I draw on several years of work with African American student researchers from disadvantaged economic backgrounds to suggest that the involvement of students as science education researchers introduces a space that is critically central to the development of student identities associated with both research and science learning.

DEFINING IDENTITY

As theorized in a 2004 study conducted with Wolff-Michael Roth, Ken-neth Tobin, Cristobal Carambo, Ya-Meer McKnight, Jen Beers, and myself, identity can be conceptualized as being constantly formed and (re)formed through our interactions in the world—identity, then, is one outcome of our activities in different spaces. Rather than a way of individually defining yourself, identity represents a recursive relationship between your own per-ceptions and other's perceptions of how you access and utilize resources. Thus, while individuals do hold a constant sense of Self over time, aspects of identity are mediated as an individual participates in social settings and utilizes various resources.

In schools, and specifically in science classrooms, a successful science learner identity is made and remade through students' interactions with (1) the community (e.g., instructor, school staff, and peers), (2) the material and human tools available (e.g., lab equipment, texts, physical classroom ar-rangements, and language), (3) the rule systems governing school practices and classroom expectations, and (4) the ways in which students are allowed to participate in class (e.g., student-centered or teacher-centered instruction). As students mediate classroom experiences, stable parts of their identity will be reinforced or shifts will take place.

If the science education community is interested in increasing students' identification with science, it is vital to consider the types of activities in which students are involved and the mediating resources that are available for their utilization. In other words, improving science learner identity formation may require providing new opportunities for student roles, community formation, rule systems, and the access and appropriation of new and existing social and material tools. I suggest that involving students as researchers of science education provides such opportunities, particularly since the activity of stu-dent research is governed by different rule systems, provides access to new and existing communities, elicits different modes of participation, and calls for the utilization of physical and social tools. While multiple outcomes can emerge when students work as researchers, amongst those outcomes are identity formation and reformation. Just how that can occur will be more closely examined in the remainder of this chapter.

STUDENT RESEARCHER ROLES

Student researchers may work during intense summer sessions or during the school year, in lunch periods or after school sessions. Regardless of when they work, their participation extends across a variety of roles as they are asked to engage in a multitude of activities. Over the past seven years, much has been learned about student researcher roles through studies conducted in conjunction with the University of Pennsylvania and several Philadelphia neighborhood schools. Importantly, we have learned that although each research site and school research team may differ in the specifics, there are benefits for involving students in roles such as interviewers, ethnographers, curriculum developers, and teacher educators. In addition to these broad roles, the division of tasks often gives the student researchers specific responsibility for data collection, artifact production, and analysis and interpretation activities. For instance, depending upon the research questions, student researchers can gather data about the curriculum, their instructor's teaching strategies, and classroom interactions between peers and between teacher and student(s). The youth may also record observations of classroom activities in journal entries, obtain views of class members through interviews, and watch videotapes to critique classroom activity. In addition, the students participate with their teacher(s) and with university-based researchers in cogenerative dialogues or conversations in which all members accept responsibility for teaching and learning practices and devise appropriate strategies for improving the classroom functionality (Tobin and Roth 2005).

LEARNING TO USE NEW AND EXISTING TOOLS IN NEW WAYS

One might wonder as to the benefits of engaging student researchers in multiple roles. In typical cases of research involvement where students may serve a member-checking role, there is limited potential to direct the research process or to make decisions about data collection, analysis, or interpretation processes. In contrast, through different and expanded roles, students have the opportunity to build new skills and to use their existing skills in new ways to interact in social settings with new and familiar groups of people. As they position themselves and are positioned as "expert" in multiple settings, the students can begin to unconsciously shift their perspectives of their range of possible actions. The following sections provide descriptions of some of the research roles that can be engaged by student researchers.

Interviewers

When students interview each other, analyses of transcripts depict that barriers dissipate between the student researcher interviewer and students being interviewed. In some cases, I have found that the students who are being interviewed begin to ask questions to the interviewers, and the interview becomes a fluid conversation. This is in stark contrast to interviews I have conducted where students were reluctant to share information and gave short responses. When student researchers act as interviewers, they can draw upon common language (e.g., slang terminology), experiences, and shared knowledge of popular symbols to connect with the students being interviewed. In this way, the student researchers provide expertise and enhanced data collection to the research group as a whole that would not have previously been possible.

Ethnographers

When student researchers act as ethnographers, they are allowed opportunities to use various media (e.g., video, photography, PowerPoint, poetry, and rap performance) to develop richly detailed representations of salient events, goals, persons, or places in their lives. Since the youth choose what is salient in their world and decide how to represent it in a way that furthers the understandings of the research group and the larger education community, they learn to use their own sets of resources (in the form of cultural knowledge and understandings of their families, neighborhoods, and schools) in new ways. Moreover, by conducting ethnographies of social spaces they attend, youth provide much needed perspectives for teachers to better understand the skills, values, and everyday realities of the students. As science teachers learn more about the lifeworld experiences of their students, they can work to connect science in more relevant ways to their students' experiences and ways of being. Classroom interactions can shift, and spaces emerge for the development of students' attraction to and confidence in science.

Teacher Educators

In addition to being ethnographers and interviewers, student researchers exhibit further expertise as they work within teacher educator roles. In this capacity, students may work closely with teacher education programs to provide much needed perspectives on science teaching and learning to prospective and practicing teachers. In small groups, student researchers provide insights, speaking about issues like what they think it means to be smart, the importance of standing up for oneself, or the characteristics of a good teacher and principal. In some cases, student researchers have joined methods courses as guest presenters to teach new teachers how to utilize video-editing technologies such as Imovie. Moreover, extending outwards from teacher education, student researchers can be involved in sharing research findings more broadly with the research community as they coauthor papers and attend professional conferences as copresenters.

Video Analysts and Theorists

One of the most important dimensions of the research process has been introducing student researchers to new forms of theoretical and canonical language as well as to analytical techniques, including video analysis. As video footage becomes an increasingly common data source in educational research, student researchers can be involved in reviewing videotapes from science classrooms to select vignettes for discussion and further analysis. For example, the youth can be asked to identify effective teaching and learning practices. The students may digitize the selected vignettes using software such as Imovie and Adobe Premier, and then export the movies into QuickTime or similar programs for use in discussion, analysis, and interpretation activities in research group meetings, cogenerative dialogues, or presentations at conferences. In the process of video vignette selection and analysis, student researchers can come to evaluate their own forms of participation in science classes. In one case, as two student researchers reviewed video footage from their chemistry laboratory, they began to discuss their chemistry lab group's engagement in the activity.

> *Monique:* And we got a good team working effort.
>
> *Dee:* We look like some young scientist right now.
>
> *Monique:* We're doin some real good observations right now. We work real good together we try to get the job done.

As students are involved in doing research on science teaching and learning, their research tasks direct their attention to their own modes of participation in the classroom, which necessarily shapes their notions of themselves as science learners. Moreover, as youth researchers are provided with theoretical lenses, they can begin to justify their selection of vignettes based upon new constructs. For instance, in my work with student researchers, they began to describe their participation in science class through sociocultural constructs such as resource access, agency, or the power to act, and building or losing social capital and status. Furthermore, they identified structures that supported their learning and examples of ways in which their participation might be limited. For example, as Monique viewed video footage, she stated,

> When we do the lab instead of us going up there to get the stuff, our stuff should already be out there.... Because it's so much confusion running back and forth to get stuff. And we start lab like at 11:00 then we only got 48 minutes and sometimes the lab take longer than that. Then we gotta do the paper. We should have more time—start earlier or something.

As student researchers work across the realms of science and research, they may begin to conflate their researcher and science learner identities. As expressed by Ivory in a journal entry following the completion of a science experiment, "[W]hatever you mix with the alka-seltzer it will exploded no matter what, but *as researchers* [emphasis added] we found out that it blows faster depending on the temperature (like hot water, coffee)."

NEW RULES: BLURRING POWER DIFFERENTIALS

When students work in new capacities, they are mediating new rule systems that position them on more level planes with adults. Student researchers have often described these new rules explicitly in public forums. For instance, Shakeem commented, "It's like we [student researchers] was in charge and they [adult researchers] was just basically sayin you've got the power . . . and we want you to come up with the full potential that you have. Just don't go crazy with the power you got." In social settings where student and adult researchers are operating under the notion that both groups have much to offer to the process, the students' identities have greater potential to shift. That is, as rules change, identity can also change and/or aspects of identity begin to blend together. This becomes evident in some of the lyrics from a rap presentation that two student researchers wrote, performed, recorded, and edited using Final Cut Pro software to capture their experiences of three years of involvement as student researchers.

> You messin' with a brand new line of researchers
> My team is nice and yours is good
> but we got the best and brightest from the worst of hoods,
> from da bridge to da fifth down to d-block
> Ol' [old] head up north. Ya'll want beef . . . not!
> Ya'll need to stop before the drama progress,
> DUS [Discovering Urban Science] stand on top above the rest
> So I guess the best what we can claim
> Puttin' Bill Nye the Science Guy and Hank to shame
> Degrees on the wall that read our name,
> Ivory, Keem, May and R is killin' the game [of life].

As Shakeem and Ivory write (on their own behalf as well as for two other student researchers in their group, May and Randy), it is evident that over time a merging of identities has resulted in a "brand new line of researchers." First, Ivory and Shakeem proudly recognize their four neighborhoods ("da bridge," "da fifth," "d-block," and "up north") and situate themselves as *researchers* within these well-established domains ("the best and brightest [researchers] from the worst of hoods"). After acknowledging stable aspects of identities associated with social spaces outside of the research setting, the student researchers reveal evidence of identity shifts, by acknowledging and uplifting their research group, Discovering Urban Science (DUS). In doing so, they establish a connection with science. They compare their group to two major figures associated with teaching children science through television, one more nationally known (*Bill Nye, the Science Guy*) and another more locally associated with the School District of Philadelphia (Tyraine Ragsdale, or Grand Hank). In doing so, the youth put forth an identification with science that positions them above these established scientists. Finally, their lyrics suggest that they not only are successful in research or science, but also anticipate success in life ("killin' the game") through their participation in these new activities. New rules for participation in science education will create spaces for marginalized youth to claim a voice in research and in science.

BUILDING COMMUNITY

> Being around people who care is a great feeling. When I was down, needed help or someone to talk to my work family was here. (Shakeem, journal entry, August 2002)

As student researchers work together, a research community is built—a "work family," as Shakeem described. Student (and adult) researchers experience changes with regard to the communities they can access and the ways in which they can connect to new groups of people. When student researchers have the opportunity to interact with other researchers who have backgrounds very different from their own, they are able to build new social networks and communicate their own values. For many of the youth with whom I have worked, having someone's back represents a deep level of social tightness. With time, reciprocal relationships of caring and loyalty can develop. For example, in our work with student researchers, my colleague, Ken Tobin, and I sometimes intervened to assist them through difficult situations (e.g., suspension); the youth extended respect and support to us as well. For instance, in another part of the rap quoted above, Ivory and Shakeem threatened imaginary adversaries, stating, "mess around with Rowhea and we'll hurtch ya" and offered thanks to Ken Tobin ("Doc"), who they described as "the light that combine us together—that give us the respect."

CODA

When students work as researchers in science education, they have the opportunity to interact in empowering ways. They seize new resources and make them their own as well as hold onto existing resources and learn to use them in new and creative manners, thus creating new opportunities for building different forms of knowledge, skill sets, and interaction patterns. With a new repertoire of resources, the students not only develop identities associated with research but also come to view themselves as individuals who can successfully participate in science class, in school, and in the larger society.

Best summarized in Ivory and Shakeem's statement "The best science squad makes great researchers," students can build aspects of identity associated with science teaching and learning in the process of becoming "great researchers." Including students in multiple researcher roles that are guided by equitable rules for interacting with new and familiar communities will contribute to their mediation of identity in multiple spaces, including the science classroom. It is time to consider new ways for conducting science education research. In working with students as researchers, we merge the worlds of research and practice. Improved classrooms and enhanced science identities can emerge not because insightful research was conducted and strategies were implemented, but rather because through the process of becoming researchers, students come to see themselves as valuable members of the community and capable science learners.

REFERENCES

Roth, Wolff-Michael, Kenneth Tobin, Rowhea Elmesky, Cristobal Carambo, Ya-Meer McKnight, and Jen Beers. 2004. Re/making Identities in the Praxis of Urban Schooling: A Cultural Historical Perspective. *Mind, Culture and Activity* 11:48–69.

Tobin, Kenneth, and Wolff-Michael Roth. 2005. Implementing Coteaching and Cogenerative Dialoguing in Urban Science Education. *School Science and Mathematics* 105:313–22.

31

Using Cogenerative Dialogues to Improve Science Education

Sarah-Kate LaVan

DIFFICULTIES OF TEACHING SCIENCE IN URBAN SCHOOLS

As members of urban school communities, we do not need to imagine the challenges our schools face in educating today's youth. We have all experienced firsthand the difficulties these schools confront in attracting and retaining qualified science teachers and the consequences the populations endure as a result; for instance, frequently changing populations, overcrowded classrooms, disconnected and low-level science curricula, and teacher-centered and -controlled classroom activities.

While studies in science education call for various remedies, I take the position that in order to support the schooling of urban students and to retain highly qualified science teachers, both teachers and students need to learn how to participate and communicate in ways that are positive and productive. Over the past five years, researchers have explored the use of cogenerative dialogues in a variety of urban schools as a means to improve many of the difficulties associated with teaching science in urban classrooms. This chapter provides an explanation of cogenerative dialogues and how these dialogues have been employed by teachers, students, and researchers to develop teacher-student interactions, transform classroom and school environments, and improve the lives of the participants.

WHAT ARE COGENERATIVE DIALOGUES?

Cogenerative dialogues are conversations in which stakeholders critically examine and reflect on shared events and activities. Although cogenerative dialogues can take many forms and serve a range of purposes, the goal of these conversations is always the same—to collectively come to an understanding about an event and to generate a future outcome. Initially, cogenerative dialogues were informal debriefing sessions held after a class between a student and one or two teachers in order to identify, discuss, and resolve problems within the classroom (Roth and Tobin 2002). These conversations allowed the participants to reflect on the day's class and activities and to positively change the classroom environment based on students' recommendations. For example, in the classrooms Tobin studied, teachers and students often used cogenerative dialogues to discuss their difficulties surrounding issues of respect and disrespect and student participation. During these conversations, participants examined the teacher's actions and jointly made suggestions about how the teacher could better show respect toward students as a way to gain greater respect and afford more productive teaching opportunities.

Recently, several teachers and researchers have found cogenerative dialogues beneficial for a variety of individual and collective purposes, and therefore have expanded these conversations to include a range of members, topics, forms, and settings. A critical component of cogenerative dialogue is that participants share responsibility for examining and creating change in the classroom. Through these dialogues, participants expose the underlying power dynamics of the school and classroom, examine people's beliefs and values, and consider a multitude of school factors and life experiences. Although some topics may seem to have a more direct impact on the teaching and learning of science, such as understanding particular classroom interactions, specific science concepts, and reasons students have difficulty completing assignments, other peripherally related topics are discussed, as they have an important influence on how participants become involved in science and interpret classroom activity and interactions.

WHERE DO COGENERATIVE DIALOGUES OCCUR?

Taking place both in and out of school settings, cogenerative dialogues occur as formal, planned meetings as well as spontaneous interactions arising from the immediate needs of particular situations. Formal cogenerative dialogues in the school setting have occurred both during and outside of class time. For example, formal cogenerative dialogues may take place between a teacher and an entire class of students during instructional periods, a teacher and student in the hallway, or a variety of stakeholders in more formal caregiver-student-teacher meetings. However, most often these dialogues take place in small-group settings between a teacher and a group of students outside of instructional time, such as immediately before or after instructional periods, during lunch periods, or after school.

Given that informal dialogues are sometimes impromptu and occur as issues arise, these conversations are not limited to within the school setting or to small-group formats. For these reasons, the numbers and types of individuals who participate in a dialogue depend highly upon the issues being addressed and the location of the discussion. For instance, an informal cogenerative dialogue that is a coplanning session between two teachers may take place either in or out of the school. However, if the coplanning session includes students, the dialogue is most likely to occur within the school when the teacher and students come into contact with one another, such as directly before or after an instructional period, in the hallway, or in the classroom. Some informal dialogues, such as huddles, may occur in the midst of teaching and learning activities. *Huddles* were originally described as interactions in which coteachers "touch base" and "fine-tune" lessons while reaching agreements about their future teaching plans, meeting teaching and learning goals, and identifying and obtaining resources (Tobin et al. 2003).

HOW HAVE COGENERATIVE DIALOGUES BEEN USED?

Changing Curricular and Classroom Structures to Support All Participants

During cogenerative dialogues, participants identify and review classroom activities, while discussing power relationships, roles of individuals, community rules, and the use of resources. Since all participants are represented, their perspectives are used to collectively inform the future actions and activities as well as aid in improving the quality of teaching. Therefore, cogenerative dialogues have become a space and time that encourage teachers and students to become actively involved in thinking about and making changes that are important to their particular classrooms and schools, rather than waiting for recommendations to be made by educators, policy makers, and researchers.

Many teachers and researchers have focused on this transformative aspect of cogenerative dialogues, focusing on their potential to directly improve individual student achievement and classroom dynamics. For instance, in her doctoral degree research Beth Wassell (2004) examined two physics coteachers' use of cogenerative dialogues as a way to explore the roles the teachers and students held and the enacted curriculum. In the cogenerative dialogues, the coteachers and students talked a great deal about the teachers' roles, the effectiveness of group work and class discussions, and suggestions for the physics curriculum. Through these dialogues, several important issues were raised, which ultimately became motivations for changing the curriculum. One significant outcome involved maintaining student interest and involvement in the subject matter. The students suggested that the teachers use innovative teaching strategies, such as skits and plays, to make the information more interesting and understandable. In cogenerating an outcome, the group decided that they could in fact employ these innovative teaching strategies, but the teachers would need help since they had never before used such activities in a science class. For these students and teachers, cogenerating this outcome supported the students' desire to make the information relevant and interesting as well as the teachers' goals of getting students to pay attention and learn the necessary material.

In addition to employing cogenerative dialogues to proactively change the classroom environment, these dialogues have been used as a means to address already apparent or troubling issues, such as concerns surrounding respect, students' failure to complete homework assignments,[1] and students' disinterest in the curricula. For example, in her doctoral degree studies Stacy Olitsky (2005) employed cogenerative dialogues to examine the implicit classroom rule structures (e.g., rules associated with taking tests, asking for help, and answering questions during class discussions) that often prevent students from succeeding in science class. Through dialogues, the physical science teacher, the students, and Stacy Olitsky not only reflected on the specific classroom structures that prevented students from demonstrating their knowledge in their eighth-grade physical science class, but also collectively agreed on concrete ways to restructure the activities to allow all students to be successful. For example, many students in the class were frequently observed turning in partially completed tests despite clearly demonstrating understanding of the material in classroom discussions and homework assignments. Through these dialogues, participants began to understand the complexity of issues surrounding students' performance on tests, including the impact of peer pressure and looking smart. Over time, the group collectively resolved this issue by creating new rules for turning in tests. For example, instead of the students turning in their papers as they finished, they had to sit quietly and draw on the back of their paper. This transformation allowed students to take their time in completing the test and not feel as though they had to turn in their tests early in order to "look smart." Seemingly a simple and inconsequential rule change, this transformation provided students the time, space, and support they needed to demonstrate their understanding, and ultimately perform better on tests. As students began to earn higher grades, their self-confidence and class participation increased.

Improving Interactions and Communication between Teachers and Students

Urban classes are typically crowded places where there is limited time and space for students and teachers to get to know one another or to discuss factors that shape teaching and learning activities. This means that teachers are often unaware of the wide range of matters that can affect student participation and achievement in their classroom. Alternatively, students are frequently unknowledgeable about the teachers with whom they interact. Across all of the sites in which cogenerative dialogues have been employed, these dialogues have provided both the teacher and students with opportunities to gain a better understanding of the other and develop a better sense of how to talk across gender, racial, and social borders.

As cogenerative dialogues are generally small in size, these discussions create an intimate space where students and teachers have opportunities to get to know one another on a personal level. By changing the power distribution within the discussions so that all stakeholders are involved in making change for the good of the class, these discussions provide all participants with a space in which to communicate their views, ideas, and beliefs about teaching, the students, and the classroom. Therefore, as teachers and students work to resolve inequitable aspects of teaching and learning, all participants take on new roles and responsibilities.

Through participation in cogenerative dialogues, students develop new ways of interacting with adults and their peers (LaVan and Beers 2005). In a cogenerative dialogue with her students, Jennifer Beers explained she felt "that one of the most important skills the students had gained, was that they were better able to advocate their views, while still showing respect." In her doctoral degree research, Sonya Martin (2005) noted that as students learned how to get their points across to others, teachers began to understand their teaching practices through others' eyes and to "truly" listen to students' concerns. Additionally, teachers learned how to explain their perspectives and experiences to students. For example, through discussions about curricular choices, such as why certain activities and homework assignments were required, students in Jennifer Beers's class came to understand that many of her practices were in fact unconscious, unintentional, and built out of frustration with the students' lack of responsibility for their learning. Conversely, through conversation Jennifer Beers began to see that students' actions too were built from their prior negative experiences with her and their interpretations of the classroom situations.

As participants explored and developed understandings about one another and the teaching and learning of science in urban classrooms, they built unity. This trust not only allowed teachers and students to sustain more discussions and make significant changes in the classroom, but also allowed participants to draw on these social ties in other spaces within the school and in discussions with others who had not been involved in cogenerative dialogues. For example, teachers often stated that their social connections with students allowed them to have an improved understanding about the students, and thus draw on these understandings in the classroom to facilitate student learning. Additionally, this better understanding of students allowed teachers to more aptly explain and advocate student perspectives to other teachers and administrators. This was observed to be particularly important for traditionally marginalized students in navigating certain aspects of the school environment.

CULTIVATING REFLECTIVE PRACTICES

The traditional structure of the urban school often places teachers in isolation, with relatively few opportunities to reflect on classroom activities with others. In rare times when teachers do have opportunities to spend time with one another, the talk typically transitions to nonreflective "teacher talk," in which complaining and gossiping occur. Cogenerative dialogues cultivate teacher reflection as the discussion focuses teachers' talk on a specific event and a future outcome. Thus, conversation shifts from "teacher talk" toward reflection and change. In a recent cogenerative dialogue, Jennifer Beers noted that while "the transition to reflecting with others and productively making change requires a conscious effort in the beginning, in time it becomes so routine that you do it constantly."

Furthermore, the structure of the school system rarely, if ever, cultivates reflective practices in students regarding their teachers' practices or their own learning. While students recognize that they perform better in some classes than others, they generally do not give much thought as to how or why this occurs. Oftentimes, when they perform poorly in science, students feel and are viewed as being deficient or incapable of learning the difficult subject matter. Cogenerative dialogue provides a time and space in which students can foster reflective skills. Over time, students gain the understanding that there are multiple factors that mediate the teaching and learning of science, such as relationships with others and the roles that students take in the classroom. Danni, a student in Sonya Martin's study, states this clearly in comparing her experiences as a sophomore in chemistry and as a junior in physics:

> I really only thought of teachers as being good or bad at teaching or me being good or bad in a subject. Now I think more about what makes the teacher good or bad or makes me perform good or bad in class. I now know there is more to it—including how teachers and students feel about one another, about the subject, and how and what they do in a class—like their actions.... [T]here are things we could look at, maybe do a different way, so we could improve it. (Danni, electronic correspondence, March 2003)

NUTS AND BOLTS OF COGENERATIVE DIALOGUES

Gaining "Buy-In"

It is important that all participants enter into the conversation with the belief that they can learn from one another and that these discussions are valuable for making change in the classroom. Therefore, "buy-in" is important to the success of the discussions. There are a few ways that may help to ensure "buy-in" in the early stages of the dialogues: (1) discuss themes that are important to all participants, such as respect, homework, or participants' home lives; (2) discuss a specific issue or event that has occurred in the classroom; or (3) show and discuss a video clip of class activity. An important part of "buy-in" is following through on making change in the classroom and addressing the needs of all participants. Therefore, it is crucial that the group either implements the cogenerated outcome or discusses why the change may not be possible. Additionally, as the group cogenerates outcomes, the participants may need to revisit topics and proposed solutions a number of times.

Rules

It is important that cogenerative dialogues are structured so that all participants feel free to speak openly and assume collective responsibility for the quality of teaching and learning in the classroom. The following rules were created to encourage participation and reduce anxiety of the discussions. Usually one or more participants begin each session by reviewing the rules and beginning the discussion.

1. No voice is privileged. *This means that no one has greater power than anyone else in the group.*
2. We must share our understandings. *Each person speaks in his or her own time.*
3. You can speak freely. *What you say does not affect your status or grade.*
4. We must cogenerate an action that benefits the individual and/or the collective.

Choosing Participants

The principle underlying these discussions is that both students and the teacher will be involved in co-constructing understandings about classroom activities and will participate in changing the classroom to make it better for all stakeholders. In an effort to gain varied perspectives, it is important to choose students who represent differing viewpoints. Typically, teachers ask students to join the group with the understanding that the involved participants will represent a cross section of the class or will provide differing perspectives about the topic of examination. For example, if the teacher is interested in examining issues surrounding achievement, it is useful to choose students from a range of achievement levels. Many studies have observed that it is best to begin these discussions with small numbers of participants. Therefore, between three and four students and one or two teachers is usually most effective.

CONCLUSIONS

As cogenerative dialogues are still relatively young, we have only just begun to examine the potential these discussions have as spaces for transforming teaching and learning. What is known, however, is that these dialogues (1) allow all participants to be involved in creating classroom change that benefits their immediate environment, (2) increase student and teacher reflection, (3) improve student-teacher understanding and interactions in and out of the classroom, and (4) allow participants to create a learning community.

NOTE

1. In Chapter 16, Jennifer Beers mentions some of the changes that resulted in her homework practices from cogenerative dialogues.

REFERENCES

LaVan, Sarah-Kate, and Jennifer Beers. 2005. The Role of Cogenerative Dialogue in Learning to Teach and Transforming Learning Environments. In *Improving Urban Science Education: New Roles for Teachers, Students and Researchers*, edited by Kenneth Tobin, Rowhea Elmesky, and Gale Seiler, 149–65. Boulder, CO: Rowman & Littlefield.

Martin, Sonya. 2005. The Social and Cultural Dimensions of Successful Teaching and Learning of Science in an Urban High School. Ph.D. diss., Curtin University of Technology, Perth, Australia.

Olitsky, Stacy. 2005. What Are the Differences in Teaching Practices and Student Learning when Science Teachers Teach Subjects That Are "Within-Field/Out-of-Field"? Ph.D. diss., University of Pennsylvania.

Roth, Wolff-Michael, and Kenneth Tobin. 2002. At the Elbows of Another: Learning to Teach through Coteaching. New York: Peter Lang.

Tobin, Kenneth, Regina Zurbano, Alison Ford, and Cristobal Carambo. 2003. Learning to Teach through Coteaching and Cogenerative Dialogue. *Cybernetics & Human Knowing* 10:51–73.

Wassell, Beth. 2004. On Becoming an Urban Teacher: Exploring Agency through the Journey of Student to First Year Practitioner. Ph.D. diss., University of Pennsylvania.

32

Distributed Leadership Practices in Science Education

Stephen M. Ritchie

In the movie *October Sky*, set in a depressing West Virginian coal-mining community during the Sputnik era of the late 1950s, the local high school science teacher (Miss Riley) encourages four male students to compete in a science fair. The boys' science fair project included the results of building and testing various rocket prototypes along with an artifact of their latest rocket design. Of course, the boys' entrance to the science fair and the progress of their project were not straightforward. Challenging community expectations and the school principal's beliefs, Miss Riley and *the Rocket Boys* triumph over numerous obstacles and eventually influence school practices—transforming their own lives as they realize their dreams.

Miss Riley was a teacher leader, and the boys' success helped the community to unite through their celebrations that in turn gave hope to fellow students. The portrayal of their aspirational interactions illustrates that leadership is relational, that it can occur anywhere within an organization, and that it is not embodied in a single designated leader like a school principal. This premise underpins what has recently become known as *distributed leadership*, where leadership tasks and practices are distributed or dispersed more widely across the organization.

Distributed leadership has become an increasingly accepted alternative perspective in the study of school leadership; it counters previous individualized leadership discourses that valorize the personal traits and actions of a single, "heroic" (typically male) principal. From a distributed perspective, leadership emerges from the interactions between members and other resources within the school or subunit of organization like the science department. Distributed leadership is not the agency of individuals but "structurally constrained conjoint agency, or the concertive labor performed by pluralities of interdependent organization members" (Woods 2004, 6). Decentering the individual leader, a distributed leadership perspective focuses on the tasks and practices that are stretched over personnel in the school or department.

Distributed leadership can manifest as teamwork. Self-selected informal teams between teachers who share ideas and resources for the development of units of work, for example, might form temporally. Alternatively, even in hierarchically structured schools, individuals like department chairs might formally convene a working party within or across the department to improve particular structures that might enhance student learning. In both cases, the human potential required for team capacity building is released and accessed as resources for/by the team. Here, teachers develop expertise by working together where the leadership that emerges collectively is more than the sum of its parts. Distributed leadership, then, empowers individuals and groups by concentrating "on engaging expertise wherever it exists within the organization rather than seeking this only through formal position or role" (Harris 2004, 13).

In this chapter I summarize what is known about distributed leadership, as well as illustrate some distributed leadership practices from recent case studies in science education. The case studies to which I refer cover a range of contexts in the United States and Australia. Finally, I suggest how schools might go about setting up structures that are more likely to foster distributed leadership practices in school science departments and classrooms.

DISTRIBUTED LEADERSHIP PRACTICES IN ELEMENTARY SCHOOLS

Spillane and his colleagues from Northwestern University (e.g., Spillane, Halverson, and Diamond 2004) are best known for their studies of distributed leadership in Chicago elementary schools. They have found that the execution of most leadership tasks involves multiple leaders and that the extent to which leadership is distributed depends on the subject area. Interestingly, leadership activity in literacy involves more leaders than in mathematics and science. More importantly, the critical question that continues to focus researcher attention in each case study is "How is leadership distributed within the school?"

They have identified three types of leadership distribution. First, *collaborative distribution* underscores the reciprocal interdependencies between individual teachers playing or feeding off one another; that is, each teacher's actions arise from interactions with other teachers that in turn fuel subsequent and continuing interactions. Similarly, collaborative distribution was the key feature of the leadership dynamics I observed in an Australian high school setting where individual science teachers felt secure to try out new practices and share them within the supportive collaborative culture of their science department (see Ritchie and Rigano 2003). Second, *coordinated distribution* refers to tasks that teachers undertake separately or together in a coordinated sequence, usually where tasks are allocated and coordinated by the designated leader. Third, *collective distribution* is leadership practice that is stretched over two or more leaders who work separately but interdependently; for example, this would be evident in co-principalships where each principal agrees on and performs his or her task responsibilities. Reference to descriptions of these practices might provide elementary teacher leaders with images of possible alternative practices relevant to their own contexts as well as raise research questions for subsequent study.

DISTRIBUTED LEADERSHIP IN HIGH SCHOOL SCIENCE DEPARTMENTS

Even though there is a growing body of literature that focuses on science teacher leaders, very few studies have been conducted within high school science departments from a distributed perspective. Starting a series of case studies of leadership within science departments from the teacher leadership literature base (see York-Barr and Duke 2004), my attention soon focused on distributed leadership, which led in turn to studying the interaction between individual and collective actions. These studies provide rich images of leadership practices that are stretched over teachers (and even students) working together in teams.

In a cross-case analysis of the leadership practices in two high school science departments in Australia (Ritchie, Mackay, and Rigano 2005), the tasks of developing and implementing units of work were the foci of the study. Even though the study was conducted in two contrasting school contexts (i.e., Saint Stephens was a Catholic girls' school, while Palm High was a regular coeducational high school that served a low-middle socioeconomic suburban community), each department depended on the collective resources produced by individual and small teams of teachers for the benefit of their respective faculty and students. The heads of departments both acknowledged the importance of drawing on these internal resources as well as accessing external resources for the purpose of improving practices within their schools. The heads exercised individual leadership roles as well as accepted influence from teachers in their departments. In this sense, the department structures enabled multiple leaders to mutually influence each other for the collective good. Yet the different school structures impacted on, but did not necessarily determine, the teaming arrangements at the two schools. Faculty at Saint Stephens experienced structures that encouraged coplanning of science units (e.g., single faculty room and teaching release for subject meetings), while the structures in place at Palm High might have supported better cross-curricular planning (e.g., science teachers were dispersed across three mixed-curricular faculty rooms) but did little to foster coplanning of science units (e.g., there was no in-school time allocated to science department meetings)—even though all relevant Palm High teachers could exercise individual agency when

it came time to implement the unit. Under these constraints, science curriculum planning was better suited to individuals working alone, but coplanning
did take place as small informal and temporal groups of teachers shared ideas
and units with each other at Palm High. While different contexts shaped the
activities of teachers, teachers created new ways to produce and contribute
collectively to the units implemented. As noted by Spillane, Halverson, and
Diamond (2001, 21), "[A]spects of the situation enable or constrain leadership
activity, while that activity can also transform aspects of the situation over
time."

ENACTING DISTRIBUTED LEADERSHIP THROUGH SHARING RESOURCES FOR COLLECTIVE USE

With my research colleagues (Ritchie et al. 2004) I explored, through
ethnography, the leadership dynamics in an academy (or school within a
school) centered on the curriculum areas of science, engineering, and
mathematics within a large urban high school in the northeastern United
States whose students were mostly African American. At the time of the
study, the academy was in transition after being formed from two previous
academies in a schoolwide restructure and where the designated leader of the
academy had just been appointed after the recent promotion of the previous
leader to assistant principal. The academy appeared to be split between two
factions, each led by a candidate for the vacated formal position of academy
leader. Loyalties were split, and there was a tendency for teachers to conduct
their work privately, in competition with each other for scarce resources,
rather than collaboratively, where resources could be shared for the collective
good. Over time, the academy became more cohesive as teachers started to
trust each other by sharing resources for collective use in the academy. These
resources were not limited to material objects; they included ideas for
teaching and management of the academy.

The academy leader accessed and distributed information about effective
teaching practices in the service of the collective interests of the academy. For
example, he recounted the successful practice used frequently by a female
teacher who successfully established a home-school partnership to a male beginning teacher who was struggling to gain respect from his students. The
teacher regularly contacted parents by telephone to inform them of the progress and achievements of her students. This helped to reinforce positive work
habits of the students at home as well as establish an effective communication
channel with the parents. By drawing on the resources available to members
of the collective in the academy, the academy leader distributed this successful
practice more widely for the benefit of individual teachers and the academy as
a whole. Successful interactions between teachers, and between teachers and
students built a sense of common purpose and belonging (or solidarity)
among members of the academy, leaving them with positive emotional energy or enthusiasm to achieve new goals.

ENACTING DISTRIBUTED LEADERSHIP THROUGH
COGENERATIVE DIALOGUE

Sharing resources and ideas for teaching and learning need not be limited to an academy leader or teachers. In the academy I studied in the northeastern United States, students also contributed to discussions that focused on improving their learning. These discussions were named *cogenerative dialogues* (see Chapter 31) because they were intended to cogenerate collective resolutions in regard to issues such as outcomes, roles, resources, and rule structures within science classrooms. Typically, cogenerative dialogues included the teacher and two or three students. They could also be used in meetings between administrative staff, parents, and their children and in whole-class settings.

In one whole-class cogenerative dialogue I observed, students were keen to suggest ways in which classroom procedures could enhance their motivation to engage in planned activities. After this meeting, both students and the teacher were committed to enacting the resolutions that were intended to improve the learning outcomes for the students and the teaching goals of the teacher. Successful outcomes from cogenerative dialogues encouraged students to exercise their collective agency in other contexts when teacher practices and academy/school structures interfered with their learning. On these occasions, aggrieved students respectfully requested participants (e.g., teacher and class) to engage in cogenerative dialogue to resolve a perceived problem. In this way, the practice of cogenerative dialogue became more widely used within the academy with greater commitment from the collective to effect agreed resolutions.

Things did not always go as planned in the academy. Students became frustrated and disrespectful in one class when a teacher failed to enact resolutions from a previous whole-class cogenerative dialogue. On this occasion, the academy leader needed to intervene to help the class renegotiate an action plan more likely to be implemented by all. It is imperative, then, that participants in a cogenerative dialogue exercise both individual and collective agency in ensuring that resolutions are achievable and that they are enacted.

From our research in the academy, we found it helpful to extend typical meanings of distributed leadership by using the more inclusive name of *collective leadership* to refer to that which involves shared responsibility of members to enact structures that afford agency to stakeholders. As well, we realized that collective leadership manifests not only as practices like cogenerative dialogues but also as solidarity among participants and the generation of positive emotional energy through successful interactions.

FOSTERING DISTRIBUTED LEADERSHIP IN
SCIENCE EDUCATION

The enactment of distributed leadership practices in schools or other organizations is likely to expand opportunities for teachers and students to participate meaningfully in the governance of those organizations. Several principles and practical suggestions for fostering distributed leadership in school science education can be gleaned from my experience in the studies outlined above.

First, a trusting and respectful climate needs to be established within school science departments to encourage the active participation of both teachers and students in generating structures that directly impinge on their daily practices. Teachers are more likely to express their goals and visions openly in such a climate, as well as share professional resources for individual and collective action and use. Students are more likely to continue contributing suggestions for improvements in teaching and learning when mutual respect between students and teacher is sealed through the enactment of previously agreed-on rules and/or practices.

Second, individual leaders like teachers and heads of departments need to orientate their professional actions toward improving outcomes for the collective by exercising agency in taking both individual and collective responsibility for enacting joint decisions, as well as creating forums for ongoing discussion.

Third, widespread use of practices like cogenerative dialogue can set up shared expectations between students and teachers within a science department; that is, students in one class who experience a cogenerative dialogue might expect to participate in a cogenerative dialogue in another class under similar circumstances.

Fourth, successful interactions between teachers, and between students and their teachers lead to solidarity and positive emotional energy. These successes are likely to create a need to experience subsequent positive interactions. Designated leaders need to harness proactive and respectful stances within departments and classrooms to ensure continued successful interactions at staff meetings and in cogenerative dialogues.

REFERENCES

Harris, Alma. 2004. Distributed Leadership and School Improvement. *Educational Management Administration & Leadership* 32 (1): 11–24.

Ritchie, Stephen M., Gail Mackay, and Donna L. Rigano. 2005. Individual and Collective Leadership in School Science Departments. *Research in Science Education* (Published online September 2, 2005).

Ritchie, Stephen M., and Donna L. Rigano. 2003. Leading by Example within a Collaborative Staff. In *Leadership and Professional Development in Science Education*, ed. John Wallace and John Loughran, 48–62. London: Routledge-Falmer.

Ritchie, Stephen M., Kenneth Tobin, Wolff-Michael Roth, and Cristobal Carambo. In press. Transforming an Academy through the Enactment of Collective Curriculum Leadership. *Journal of Curriculum Studies*.

Spillane, James P., Richard Halverson, and John B. Diamond. 2001. Investigating School Leadership Practice: A Distributed Perspective. *Educational Researcher* 30 (3): 23–28.

————. 2004. Towards a Theory of Leadership Practice: A Distributed Perspective. *Journal of Curriculum Studies* 36:3–34.

Woods, Philip A. 2004. Democratic Leadership: Drawing Distinctions with Distributed Leadership. *International Journal of Leadership in Education* 7 (1): 3–26.

York-Barr, Jennifer, and Karen Duke. 2004. What Do We Know about Teacher Leadership? Findings from Two Decades of Scholarship. *Review of Educational Research* 74:255–316.

SUGGESTED READINGS

Lingard, Bob, Debra Hayes, Martin Mills, and Pam Christie. 2003. *Leading Learning. Making Hope Practical in Schools.* Maidenhead: Open University Press.

Spillane, James P. 2005. *Distributed Leadership.* San Francisco: Jossey-Bass.

WEBSITE

Details of the study and publications from James Spillane and his colleagues at Northwestern University: www.letus.org/dls/index.htm